NUTRIENTS AND GENE EXPRESSION

Clinical Aspects

NUTRIENTS AND GENE EXPRESSION

Clinical Aspects

Edited by

CAROLYN D. BERDANIER

CRC Press
Boca Raton New York London Tokyo

Acquiring Editor: Robert A. Stern
Project Editor: Erica Orloff
Marketing Manager: Becky McEldowney
Cover Design: Dawn Boyd
PrePress: Kevin Luong
Manufacturing: Sheri Schwartz

Library of Congress Cataloging-in-Publication Data

Nutrients and gene expression : clinical aspects / edited Carolyn D. Berdanier.
p. cm. -- (Modern nutrition)
Includes bibliographical references and index.
ISBN 0-8493-9485-6 (alk. paper)
1. Nutritionally induced diseases--Genetic aspects. 2. Gene expression. I. Berdanier, Carolyn D. II. Series: Modern nutrition (Boca Raton, Fla.)
RC622.N86 1996
616.3′9042--dc20 96-14781
CIP

International Standard Book Number 0-8493-9485-6
Library of Congress Card Number 96-14781
Printed in the United States of America 1 2 3 4 5 6 7 8 9 0
Printed on acid-free paper

MODERN NUTRITION

Edited by Ira Wolinsky and James F. Hickson, Jr.

Published Titles

Manganese in Health and Disease, Dorothy Klimis-Tavantzis
Nutrition and AIDS: Effects and Treatment, Ronald R. Watson
Calcium and Phosphorus in Health and Disease, John B. Anderson and Sanford Garner
Nutrition Care for HIV-Positive Persons: A Manual for Individuals and Their Caregivers, Saroj M. Bahl and James F. Hickson, Jr.

Forthcoming Titles

Nutrition and Health: An International Perspective, Saroj M. Bahl
Zinc in Health and Disease, Mary E. Mohs

Edited by Ira Wolinsky

Published Titles

Practical Handbook of Nutrition in Clinical Practice, Donald F. Kirby and Stanley J. Dudrick
Childhood Nutrition, Fima Lifshitz
Nutrition and Health: Topics and Controversies, Felix Bronner
Nutrition and Cancer Prevention, Ronald R. Watson and Siraj I. Mufti
Advanced Nutrition: Macronutrients, Carolyn D. Berdanier
Handbook of Dairy Foods and Nutrition, Gregory D. Miller, Judith K. Jarvis, and Lois D. McBean
Nutritional Concerns of Women, Ira Wolinsky and Dorothy Klimis-Tavantzis

Forthcoming Titles

Laboratory Tests for the Assessment of Nutritional Status, 2nd Edition, H. E. Sauberlich
Nutrition and Hypertension, Michael B. Zemel
Nutrition: Chemistry and Biology, 2nd Edition, Julian E. Spallholz and L. Mallory Boylan
Advanced Nutrition: Micronutrients, Carolyn D. Berdanier
Child Nutrition: An International Perspective, Noel W. Solomons
Antioxidants and Disease Prevention, Harinder S. Garewal

SERIES PREFACE FOR CRC SERIES IN MODERN NUTRITION

The CRC Series in Modern Nutrition is dedicated to providing the widest possible coverage to topics in nutrition. Nutrition is an interdisciplinary, interprofessional field par excellence. It is noted by its broad range and diversity. We trust that the titles and authorship in this series will reflect that range and diversity.

Published for a broad audience, the volumes of the CRC Series in Modern Nutrition are designed to explain, review, and explore present knowledge and recent trends, developments, and advances in nutrition. As such, they appeal to professionals, as well as, educated laymen. The format for the series will vary with the needs of the author and the topic, including, but not limited to, edited volumes, monographs, handbooks, and texts.

Contributors from any bona fide area of nutrition, including the controversial, are welcome.

We welcome to the Series *Nutrients and Gene Expression: Clinical Aspects*. This volume is a companion volume to *Nutrition and Gene Expression* by C. D. Berdanier and L. L. Hargrove (Eds.) also published by CRC Press.

Ira Wolinsky, Ph.D.
Series Editor

THE EDITOR

Carolyn D. Berdanier, Ph.D., is a Professor of Nutrition at the University of Georgia in Athens, Georgia. She received a B.S. degree from The Pennsylvania State University and M.S. and Ph.D. degrees in Nutrition from Rutgers University in 1966. After a post-doctoral fellowship year with Dr. Paul Griminger at Rutgers, she served as a Research Nutritionist with the Human Nutrition Institute which is a part of ARS, a unit of the U.S. Department of Agriculture. In 1975 she moved to the University of Nebraska College of Medicine where she continued her research in nutrient gene interactions. In 1977 she moved to the University of Georgia where she served as Head of the Department of Foods and Nutrition. She stepped down from this post ten years later and devoted her full time efforts to research and teaching in her research area. Her research on the diet and genetic components of diabetes and vascular disease has been supported by NIH, USDA, U.S. Department of Commerce, The National Livestock and Meat Board, and the Egg Board. She is a member of the American Institute of Nutrition, the American Society for Clinical Nutrition, The Society for Experimental Biology and Medicine, and several honorary societies in science. She has served on the Editorial Boards of the *FASEB Journal, The Journal of Nutrition,* and *Nutrition Research and Biochemistry Archives*. She is also a Contributing Editor for *Nutrition Reviews* and Editor of the AIN News Notes. Current research interests include studies on aging, the role of diet in damage to mitochondrial DNA, and the role of specific dietary ingredients in the secondary complications of diabetes.

PREFACE

Man is subject to a wide variety of lethal and non-lethal diseases. The intensity of the disease state as well as its duration is a function of an interaction between the genetically determined resistance to a particular disease and a host of environmental factors of which nutrition is but a part. Even a simple, communicable disease due to a known infectious agent such as measles or chicken pox can be highly variable with respect to symptoms and degree of incapacitation. This variation can be understood through an understanding of the sometimes complex interaction of genes that encode the elements of the immune system and specific factors which turn on or turn off the expression of these genes. Some of these factors can be nutrients, while others may be hormones and still others may be specific metabolites or specific minerals or nearby genes.

In the first volume of this series on nutrition and gene expression, individual nutrients were discussed with respect to their effects on gene transcription and translation. Individual genes important to metabolism were also discussed with respect to their regulation and function in nutritionally important processes. The present volume addresses a group of genetic diseases which currently are believed to develop as a result of nutrient–gene interactions. Diabetes mellitus in its many forms, cardiovascular disease as a function of disturbances in genes which encode lipoproteins and their receptors, obesity, aging, programmed cell death and disturbances in branched chain amino acid metabolism are all discussed from the genetic–nutrient interaction point of view. As the reader proceeds through the book, one hopes that the imagination and dedication of all the many scientists whose reports were used to construct these reviews will be appreciated. Without their scientific expertise this book would not have been possible. To them I give my admiration and gratitude.

Carolyn D. Berdanier

CONTRIBUTORS

Carolyn D. Berdanier, Ph.D.
Department of Foods and Nutrition
University of Georgia
Athens, Georgia

Claude Bouchard, Ph.D.
Physical Activity Sciences
Laboratory
Laval University
Ste-Foy, Québec, Canada

Pamela J. Fraker, Ph.D.
Department of Biochemistry
Michigan State University
East Lansing, Michigan

Bin He, M.S.
Department of Foods and Nutrition
University of Georgia
Athens, Georgia

Nobuko Iritani, Ph.D.
Tezukayama Gakuin College
Osaka, Japan

Shu-Huei Kao, M.S.
Department of Biochemistry
School of Life Science
National Yang-Ming University
Taipei, Taiwan

Roy J. Martin, Ph.D.
Department of Foods and Nutrition
University of Georgia
Athens, Georgia

Harbhajan S. Paul, Ph.D.
Biomed Research and Technologies,
Inc.
Wexford, Pennsylvania

Marty H. Porter, M.S.
Institute f. Nutztierwiss
ETH-Zentrum
Zurich, Switzerland

William Telford, Ph.D.
Medical Sciences
University of Michigan
Ann Arbor, Michigan

Yau-Huei Wei, Ph.D.
Department of Biochemistry
School of Life Science
National Yang-Ming University
Taipei, Taiwan

B. Douglas White, Ph.D.
Department of Foods and Nutrition
University of Georgia
Athens, Georgia

TABLE OF CONTENTS

Chapter 1

NUTRITION AND GENETIC DISEASES

Carolyn D. Berdanier

CONTENTS

INTRODUCTION

In the first volume[1] of this series on nutrition and gene expression, a variety of topics were presented which addressed the issue of how the nutritional state or specific nutrients affected gene expression in eukaryotic cells. In this volume, the theme is reversed. That is, the topics concern genetic diseases or genetically controlled processes on which nutrients have specific effects, vis-a-vis the disease process. Our knowledge in many instances is incomplete. We might realize that a given disease or a tendency to develop that disease is inherited, but we do not know which gene(s) are responsible, nor, if we have a candidate gene, exactly how specific nutrients affect its expression. In the first volume, genetically dictated events in the body were described that could be modified by the nutritional state of the individual. In this volume, genetic diseaes are discussed not as disease entities per se but as metabolic states that are or may be the result of one or more nutrient–gene interactions. Such interactions are an essential component of life. Indeed, man's survival through the ages is the result of an interaction between his nutrient intake and his genetic adaptability.

Survival in an environment which constantly changes is optimal in that individual who can sense these changes and respond to them. Food as a part of the environment provides both the challenge to the body to obtain it and to use

0-8493-9485-6/96/$0.00+$.50

it to sustain normal health, growth, development, and lifespan. Both the quantity and the composition of the available food can be highly variable. The key to survival therefore, rests with the genetic heritage of the individual that in turn provides the ability to optimally use the food consumed. Implicit in this understanding of survival under adverse circumstances is an understanding of how genes are expressed and how the products of this gene expression function under a variety of circumstances. In addition, an understanding must develop of the consequences of aberrant gene expression and how specific nutrients can affect this expression. The scope of this volume addresses diseases that develop as a consequence of aberrant genes and/or which may be explained by an interaction of nutrients and genes. Several clinical situations have been selected for presentation to provide an appreciation for the breadth of disorders which can develop as a result of one or more mutations and which are responsive to one or more nutrients. Single gene defects as well as diseases of multiple origin are included.

SINGLE GENE DEFECTS

Over time, evolution of the species, homo sapiens, has involved a multitude of changes in the genetic material, DNA. Those changes which improved the chances of survival of the species were passed on to subsequent generations. Those which impaired survival usually were not carried forward. Today, genetic aberrations (mutations) that result in discernible disease can, in some instances, be diagnosed early and in some instances, managed such that early death is avoided. This is not always possible. Listed in Table 1 are diseases attributable to one or more mutations in a single gene. This is not an all-inclusive list but is merely a sampling of this type of disease. The diseases listed in this table are those which have resulted from mutations in the nuclear genome. In some instances, the exact mutation and its chromosomal location are known. In others, the defect is known, but the location of its corresponding gene as well as its sequence is not known. The nuclear DNA is being sequenced and mapped (The Human Genome Project), but the particular attributes and features of aberrant nuclear genes are not always available. Thus, even though we might know that an enzyme or carrier or receptor that is aberrant due to a mutation in the nuclear genome, we might not know whether this mutation was a point mutation (a substitution of bases in the chain) or a deletion mutation (one or more bases deleted). Further, we might not even know the normal base sequence for the gene in question or its chromosomal location. In a number of genetic diseases we might only know its mode of inheritance. In other words, much more information about the nuclear genome is needed before an understanding of its role in specific human disease will be in hand.

MITOCHONDRIAL DISEASES

In contrast to the nuclear DNA, mitochondrial DNA (mtDNA) has been completely sequenced and mapped. Mutations in this DNA have been described and resultant diseases compiled (Table 2).[2] Human mtDNA is about

TABLE 1
Diseases Attributable to One or More Mutations in a Single Gene

Disease	Mutated Gene Product	Characteristics
Tay-Sachs Disease	Hexosaminidase A	Early death, CNS degeneration, ganglioside GM_2 accumulates
Gaucher's Disease	Glucocerebrosidase	Enlarged liver and spleen, erosion of long bones and pelvis, mental retardation, glucocerebroside accumulates
Fabry's Disease	α Galactosidase A	Skin rash, kidney failure, pain in legs and feet, ceramide trihexoside accumulates
Niemann-Pick Disease	Sphingomyelinase	Enlarged liver and spleen, mental retardation, sphingomyelin accumulates
Krabbe's Disease (Globoid Leukodystrophy)	Galactocerebrosidease	Mental retardation, absence of myelin
Metachromatic Leukodystrophy	Arylsulfatase A	Mental retardation, sulfatides accumulate
Generalized gangliodosis	Gmi, Ganglioside: β galactosidase	Mental retardation, enlarged liver
Sandhoff-Jatzkewitz Disease	Hexosaminidase A and B	Same as Tay-Sachs but develops quicker
Fucosidosis	α-1-Fucosidase	Cerebral degeneration, spastic muscles, thick skin
Refsum's Disease	α Hydroxylating enzyme	Neurological problems: deafness, blindness, cerebellar ataxia, phytanic acid accumulates
Albinism	Tyrosinase	Lack of melanin (skin pigment) formation; increased sensitivity to sunlight, lack of eye pigment
Alcaptonuria	Homogentisate oxidase	Elevated urine levels of homogentisate; slow deposits of homogentisate in bones, connective tissue and internal organs resulting in gradual darkening of these structures. Increased susceptibility to arthritis.
Parkinson's disease	Enzyme not identified	Decreased dopamine production by certain brain areas resulting in muscle tremors
Histidenemia	Histidase	Elevated levels of histidine in blood and urine. Can give false positive result in test for phenylketonuria. Elevated urocanase levels in sweat. Decreased histamine formation.
Maple syrup urine disease	Branched chain keto acid dehydrogenase (Several variants) ketoacidosis, early death	Elevated levels of a ketoacids and their metabolites in blood and urine; mental retardation,

TABLE 1 (CONTINUED)

Disease	Mutated Gene Product	Characteristics
Methylmalonuria	Methylmalonyl CoA mutase (several variants)	High blood levels of methylmalonate; Inability to use vitamin B_{12}, pernicious anemia, early death
Nonketotic hyperglycinemia	Glycine cleavage enzyme	Severe mental retardation, early death, high blood glycine levels
Hypermethioninemia	Methinoine adenosyl-transferase (↑ Km not deficiency per se)	Accumulation of methionine in blood (condition is benign)
Homocysteinemia	Cystathionine synthase	Elevated levels of cystathionine in urine (condition in benign)
Phenylketonuria	Phenylalanine hydroxylase (several variants)	Increased levels of phenylalanine and related metabolites in blood and urine; mental retardation; decreased neurotransmitter synthesis; shortened lifespan
Tyrosinemia	Tyrosine transaminase	Eye and skin lesions, mental retardation
	Fumarylacetoacetate hydrolase	Liver failure, renal failure
	p-Hydroxylphenylpyruvate oxidase	Increased need for ascorbic acid
Lactose intolerance	Lactase	Chronic or intermittent diarrhea, flatulence, nausea, vomiting, growth failure in young children
Sucrose intolerance	Sucrase	Diarrhea, flatulence, nausea, poor growth in infants
Glucose-galactose intolerance	Glucose-galactose carrier	Diarrhea, growth failure in infants, stools contain large quantities of glucose and lactic acid
Galactosemia	Galactose 1-P-uridyl	Increased cellular content of galactose 1-phosphate, eye cataracts, mental retardation, increased cellular levels of galactitol
	Galactokinase	Cataracts, cellular accumulation of galactose and galactitrol
	Galactopimerase	No severe symptoms
Fructosemia	Fructokinase	Fructosuria, fructosemia
	Fructose 1-P aldolase	Hypoglycemia, vomiting after fructose load, fructosemia, fructosuria; in children: poor growth, jaundice, hyperbilirubinemia, albuminuria, amino-aciduria
	Fructose 1,6-diphosphase	Hypoglycemia, hepatomegaly, poor muscle tone, increased blood lactate levels

Pentosuria	NADP-lined xylitol dehydrogenase	Elevated levels of xylose in urine
Hemolytic anemia	Glucose-6-phosphate dehydrogenase	Low erythrocyte levels of NADPH, hemolysis of the erythrocyte
Type VII glycogenosis	Phosphofructokinase	Intolerance to exercise, elevated muscle glycogen levels, accumulation of hexose monophosphates in muscle
Von Gierke's Disease (Type I glycogenosis)	Glucose-6-phosphase	Hypoglycemia, hyperlipemia, brain damage in some patients, excess liver gycogen levels, shortened lifespan, increased glycerol utilization
Amylopectinosis (Type IV glycogenosis)	Branching enzyme Liver amylot (1,4 →1,6)-transglucosidase	Tissue accumulation of long-chain glycogen that is poorly branched, intolerance to exercise
Pompe's disease (Type II glycogenosis)	Lysosomal α-1,4-glucosidase (acid matase) Amylo-1,6-glucosidase (Debranching enzyme)	Generalized glycogen excess in viscera, muscles, and nervous system, extreme muscular weakness, hepatomegaly, enlarged heart
Forbe's disease (Type III glycogenosis)	Muscle phosphorylase	Tissue accumulation of highly branched, short-chain glycogen, hypoglycemia, acidosis, muscular weakness, enlarged heart
McArdle's disease (Type V glycogenosis)	Liver phosphorylase	Intolerance to exercise

16,569 base pairs (bp) in length.[3] This DNA codes for 22 tRNAs, 2 ribosomal RNAs and 13 components of oxidative phosphorylation (OXPHOS).[2-8] Less than 10% of the genome is noncoding. Of this 10%, 90% is found in the 1,122-bp control unit between the proline and phenylalanine tRNA sequences. This control unit contains the origin of heavy strand replication,[3] the origins of both heavy and light strand transcription,[4] promoters for both heavy and light transcription,[5,7] two transcription factor binding sites, three conserved sequence blocks associated with the initiation of replication, and the D-loop strand termination associated sequences.[7] The mtDNA codes for seven proteins of Complex I, one protein in Complex III, three proteins in Complex IV of the respiratory chain, and two proteins in F_0 portion of Complex V, the F_1F_0ATPase.

Nuclear encoded proteins play important roles in mitogenesis. Some of these are enzymes that are needed for the synthesis of the components of oxidative phosphorylation while other nuclear genes have functions that relate to the expression of individual mitochondrial gene products such as cytochrome b and the subunits 1 and 2 of cytochrome oxidase.[5,7,8]

The fact that genetic information is distributed among two spatially separate compartments implies the existence of some mechanism for ensuring a coordinated expression of the proteins and/or RNAs encoded in the two genomes. Growth and division of eukaryotic cells is accompanied by a concomitant increase in mitochondrial mass. This is accomplished not only by the *de novo*

TABLE 2
Mitochondrial Mutations and Resultant Disorders

Disorder	Mutation Location	Disease Characteristics
MELAS	Complex IV, Cytochrome c Oxidase	Mitochondrial myopathy, encephalopathy, lactic acidosis, stroke-like episodes, early death
MELAS	$tRNA^{Leu(UUR)}$	
MELAS	T to C at 3271	
MELAS	A to G at 3243	
MELAS	ND4 Subunit of Complex 1	Subacute necrotizing encephalomyelopathy, psycyhomotor retardation, feeding and respiration abnormalities, early death
Leigh Syndrome	Complex I, NADH:Q1 oxidoreductase	
Leigh Syndrome	Complex IV, Cytochrome c Oxidase	
MERRF	$tRNA^{lys}$, small deletions in Complex I and IV	Myoclonic epilepsy associated with ragged-red muscle fibers
IDDM	10.4 kb Duplication deletion in light strand origin of mtDNA replication	Insulin dependent diabetes mellitus
NIDDM	G to A at bp 8852, F_0ATPase $tRNA^{Leu(UUR)}$	Maturity onset, mild to moderate non insulin dependent diabetes mellitus
NARP	T to G at bp 8993 of ATPase 6	Neurogenic muscle weakness, ataxia, retinitis pigmentosa, seizures, blindness, dementia
LHON	T to C at 3394, ND1	
LHON	G to C at 3460, ND1	
LHON	T to C at 4160, ND1	Acute or subacute visual loss caused by severe bilateral optic atrophy and large centrocecal scotomata
LHON	T to C at 4216, ND1	
LHON	A to G at 4917, ND2	
LHON	G to A at 5244, ND2	
LHON	G to A substitution in ND4	
LHON	G to A at 7444, CO1	
LHON	G to A at 13708, ND5	
LHON	T to C at 14484, ND6	
LHON	G to A at 15257, Cyt b	

formation of the organelle, but also by the addition of lipid and newly synthesized proteins to the preexisting mitochondria, thus ensuring constancy of function. The latter condition depends on the balanced rates of synthesis of the catalytic and structural constituents by both the cytoplasmic and mitochondrial systems. In the five complexes of oxidative phosphorylation, there are four that have some of their subunits encoded by the nuclear genome and some by the mitochondrial genome. Each of these groups of proteins interacts to comprise the intact system for oxidative phosphorylation.

Listed in Table 2 are diseases attributed to mutations in the mtDNA. Most of these are relatively rare. As with any genetic disease, the severity and time course of development is determined by the location of the mutation within a specific gene. However, in the instance of mtDNA mutation, there is a further consideration. In contrast to diseases caused by mutations in nuclear DNA, mutations in mtDNA might not be expressed. This is because there are many mitochondria in each cell and not all these mitochondria carry the same DNA. Some might be normal while others might be mutated. A liver cell, for example, may have as many as 800 mitochondria; an oocyte may have many thousand. Heteroplasmy can occur in these cells where some of the mitochondria have a normal base sequence in their genome while others have a mutated sequence. Homoplasmic cells are those containing mtDNA with identical sequences. Two types of mutations can occur; in one, several bases are missing or are deleted. In the other, there is a substitution of a base in the code which in turn results in an amino acid substitution in the translation product. Depending on where the base pair substitution or deletion occurs with respect to the active site of the translation product, the effect of the mutation could be minimal, modest, or profound with respect to the function of that particular mitochondrion. However, because there are many mitochondria in each cell, the overall effect could be minimal. Whether a clinically recognized disease occurs would depend on the proportion of mutant to normal mitochondria in each cell and tissue. In addition, there can be individual variation in the degree of heteroplasmy. Even identical twins can differ in the percent mutated DNA within their cells. Within the same individual there can be tissue differences as well. For example, fibroblasts could have more mutated DNA than liver cells, while muscle cells could have very little mutated mtDNA. The reverse can also occur. As mentioned, there are many mitochondria in any given cell, so if a small percentage are aberrant, the existence of the mutation might not be known. However, should a large percentage be aberrant, there could be measurable consequences. The threshold for detection of a phenotypic expression of mutated mtDNA is tissue-dependent and dependent on the location of mutation in the genome. Shoubridge[10] has pointed out that a translation product defect is not expressed until the ratio of mutant to normal mtDNA exceeds a specific threshold. Below this threshold the phenotype can be rescued by intramitochondrial complementation. In those diseases characterized by skeletal muscle pathology, the typical mosaic expression of this pathology is determined by both the cellular and organellar distribution of the mtDNA mutants.

Just as cells can be heteroplasmic with respect to mixtures of mtDNA with normal and substituted sequences, so too can cells be heteroplasmic with respect to mtDNA of different lengths. However, in the instance of the heteroplasmic cell with the deleted mutation, there is the tendency to drift towards deletion mutation homoplasmy. The reason for this drift is because mitochondria reproduce themselves at a rate that is five to ten times faster than the rate of cell replication and shorter strands of DNA are replicated at a faster rate than strands of normal length.[8] Mutations in the mtDNA which allow for survival until the fourth, fifth or even sixth decade of life i.e., diabetes (see Chapter 8) or premature aging (Chapter 9) are contributed to the gene pool via maternal inheritance.[9] This is because at fertilization almost none of the sperm's mitochondria enter the egg. Thus, mtDNA mutation(s) in the father are not transmitted to the progeny while those of the mother are or could be.

Both large and small deletion mutations have been reported.[2,10-13] Deletion mutations could occur due to slipped mispairing between repeated sequences during DNA replication.[11,12] Domains containing tandemly repeated DNA sequences are often highly polymorphic in length due to the propensity of repeat units to undergo addition or deletion events. Slipped mispairing between adjacent or nearby repeat sequences during replication is one of several proposed mechanisms for mtDNA deletion.[11]

Slipped mispairing between distant repeats can also occur and may be responsible for larger scale deletions associated with a variety of neuromuscular diseases in humans.[12-15] Deletions can also occur as a result of free radical attack on the genome (see Chapter 9). Age-dependent deletion has been reported in human tissue,[15-18] and these deletions have been found to parallel an age-related decline in respiratory function.[19] Because certain genomes may have more areas of direct repeats than others and because deletions are more likely to occur in or near these areas, these genomes might be more vulnerable to this type of mutation.[2,10,13,14]

Recently, a patient with Pearson's syndrome (a multisystem failure involving the hematopoeitic system, the pancreas, liver, and kidney) has been described.[15] He had a deletion mutation in the ND1 region of varying prevalence in the body. The liver was the most profoundly affected, having 96% of its mtDNA in the deleted form. The kidney had 87% of its mtDNA in the deleted form while the muscle had no detectable mutation at all. The patient died in his third year of life. As this study showed, a deletion mutation can become evident faster than a point mutation. As the animal or human ages, the drift towards deletion mutation homoplasmy accelerates and several investigators have reported that age is a critical factor in the detection and accumulation of these deletions.[16-22]

Mutations in the codes for the tRNAs have been reported[23-26] as well as mutations in the controller regions.[27] Controller region sequence mutation has been reported to occur at rates faster than elsewhere in the genome, and the sites of these mutations are distributed along the region rather than being clustered.[27] Polymorphic variation occurs as well with no untoward effects on the gene products.[27,28]

Documented mutations in complexes I, III, and IV of the respiratory chain are characterized by blindness or epilepsy or neurological dysfunction. The syndrome MELAS, characterized by early onset of mitochondrial myopathy, encephalopathy, lactic acidosis, and stroke-like episodes, can arise from a mutation of the mtDNA gene for cytochrome c oxidase. Several mutations have been identified but the phenotypic expression of these genotypes is similar.[29-36] There is a marked reduction in cytochrome c oxidase activity with the result of a marked decrease in oxygen use by the respiratory chain and a marked reduction in ATP production. MELAS also results from a point mutation in the ND4 subunit of the OXPHOS complex I[34] and from a point mutation in the tRNA for leucine,[35] and from a general loss in mtDNA in muscle or liver, or both.[36] In each of these instances, lifespan is reduced. Early death also characterizes children with Leigh syndrome,[37] in which mutations in the genes for cytochrome c oxidase and NADH:Q, oxidoreductase have been found. These mutations occur in different locations from those found in patients with MELAS.

In patients with a form of epilepsy known as ragged red fiber disease or myoclonic epilepsy associated with ragged red muscle fibers (MERRF), several defects have been found.[38-44] A mutation in the tRNA for lysine and small deletions in the genes for complex I and IV have been reported. Having a mutation in any of the tRNAs for any of the amino acids could have large effects on the functional characteristics of the translation products of all of the mitochondrial genes. This is particularly true for those amino acids which provide reactive groups (SH groups or terminal hydroxyl or amino or carboxyl groups) that are part of the active site of the particular protein. In MERRF it appears that the effects of the point mutation in the tRNA for lysine and the deletion mutations in complex I and IV are additive and lethal. In such instances where only one of these mutations exists, the disease is somewhat milder but nonetheless eventually lethal.

Leber's hereditary optic neuroretinopathy (LHON) is one of the more prevalent of the mitochondrial diseases.[45] People with this disorder develop blindness and 75% of the victims are female. The disease is due to a base substitution at position 11778 which changes a G to an A and which results in the substitution of histidine for arginine in one of the seven subunits of NADH-ubiquinone oxidoreductase. LHON has also been attributed to several other point mutations in the ND molecule.[43-45] Lastly, noninsulin-dependent diabetes mellitus and insulin-dependent diabetes have been attributed to mutations in the mtDNA. These are discussed in Chapter 8.

Single gene defects and subsequent disease can arise through defects in the control of gene expression as well as through point and deletion mutations in the specific genes. Thus, a review of this control is in order. This will be followed by an example of a disease (maple syrup urine disease) in which defects in the gene and its control have been identified. Chapter 2 will continue this discussion by describing studies on the control of the metabolism of branched-chain amino acids.

CONTROL OF GENE EXPRESSION

In a complex mammal such as man, there is considerable diversity in cell function. The neural cell, the hepatocyte, and the adipocyte all have very specific characteristics, yet the nuclei in these cells carry identical DNA. It is identical in length and sequence, yet not all of the genes in this DNA are transcribed and translated into mature gene products. This implies that the process is highly controlled. Only those proteins needed by a specific cell for a specific purpose are synthesized. Research has shown that gene expression is regulated at each step in the pathway from DNA to RNA to protein. This includes transcription control, RNA processing control, RNA transport control, translational control, mRNA stability control, and post-translation control. Each of these control points is nutritionally responsive in some fashion. For most genes, transcriptional control is, perhaps, the strongest control point.[46] This control is exerted by that portion of the DNA called the promoter region. This region is responsible for binding the enzyme, RNA polymerase II, plus a number of transcription factors. Because promoters are located in the 5′ flanking region just prior to the structural gene on the same strand of DNA, they are referred to as cis-acting elements. Trans-acting elements are those factors (usually proteins) produced by other genes which influence transcription. Trans-acting elements could be protein hormones such as insulin or a steroid hormone–receptor protein complex.

The promoter usually contains the start site for RNA synthesis. When the RNA polymerase binds tightly to this specific DNA sequence, RNA transcription begins. The RNA polymerase opens up a local region of the DNA double helix so that the gene to be transcribed is exposed. One of the two DNA strands acts as the template for complementary base pairing with incoming ribonucleoside triphosphate monomers. The nucleotides are joined until the polymerase encounters a special sequence in the DNA called the termination sequence. At this point transcription is complete and the mRNA is then available for translation. This outline of transcription has omitted a number of important details with respect to transcription control. For example, regulation of transcription is exerted by a group of proteins that determines which region of the DNA is to be transcribed. The cells contain a variety of sequence-specific DNA binding proteins. These proteins are of low abundance and they function by binding to specific regions on the DNA. These regions are variable in size but are usually between 8 and 15 nucleotides. Depending on the binding protein, transcription is either enhanced or inhibited and indeed cell types may differ because of these proteins. Since all cells contain the same DNA, its expression can be controlled at this point simply by the binding of these very specific DNA binding proteins.[48] Thus, genes for the enzymes of gluconeogenesis, for example, could be turned on in the hepatocyte but not in the myocyte or adipocyte simply because the hepatocyte has the needed specific DNA binding protein that the other cell types lack. At some point in differentiation, these cells failed to acquire sufficient amounts of these proteins so as to have the gluconeogenic capacity of the hepatocyte.

Actually, specific DNA binding proteins as regulators of gene expression comprise only a part of this regulation. Most genes are regulated by a combination of regulatory elements.[48] In some, a group of DNA binding proteins interact to control the activation and inhibition of transcription. Not all of these proteins are of equal power in all instances. There may be a "master" regulatory protein that serves to coordinate the binding of several "lesser" proteins.[47] This is important for the coordinate expression of genes in a single pathway. For example, in the lipogenic pathway (see Chapter 7) the genes which encode the enzymes of the fatty acid synthase complex are coordinately expressed. That is, whether increased or decreased, all enzymes are synthesized to the same extent so that the pathway can function efficiently. Another example is the series of genes which encode the enzymes needed for the conversion of a fibroblast to a myocyte. The mammalian skeletal muscle cell is very large and multinucleated. It is formed by the fusion of myoblasts (myocyte precursor cells) and contains characteristic structural proteins, as well as a number of other proteins that function in energy metabolism and nerve–muscle signaling. When muscle is being synthesized, all of these proteins must be synthesized at the same time. In proliferating myoblasts, very few of these proteins are present yet as these myoblasts fuse, the messenger RNAs for these proteins increase as does the synthesis of the proteins. This indicates that the expression of the genes for muscle protein synthesis is coordinated and controlled at the level of transcription and also, that this process is responding to a single gene regulatory DNA binding protein. This protein (myo D1) has been isolated and identified and occurs only in muscle cells. Should this protein be inserted in some other cell type, a skin cell or an adipocyte, the same expression will occur. That is, the skin or fat cell will look like a muscle cell. It will take on the characteristics of a myoblast and become a myocyte. Of interest is the fact that although all of the genes needed for synthesis in the myocyte and its master controller are present, synthesis will not occur or will occur at a very limited rate when one or more of the essential amino acids needed for this synthesis are absent or deficient in the diet. Here is an example of a gene–nutrient interaction that has control properties with respect to muscle protein synthesis and will ultimately affect the overall process of growth. Turning this situation around, if the master regulator, myo D1, is aberrant or if one or more of the genes which encode the enzymes needed for protein synthesis in the myocyte have mutated such that the enzyme in question is nonfunctional or only partly functional, muscle development will cease or be retarded. In either instance abnormal growth will result.

As mentioned, transcription is regulated by both the nearby upstream promoter region and the distant enhancer elements. The upstream enhancer element includes a TATA box and extends for about 100 bp. Enhancer fragments further upstream can bind multiple proteins which in turn can influence transcription. These elements are proteins and are labeled jun, AP2, ATF or CREB, SP1, OTF1, CTF or NF1, SRF and others. One well-studied group of proteins are those which also bind steroid hormones. These are called the steroid response elements. Steroids that enhance (or impair) transcription act by binding to one of these specific proteins and comprise this steroid response element.[49] These elements thus enable cells to respond to a steroid hormone stimulus. This element has been well studied and consists of 100 amino acids

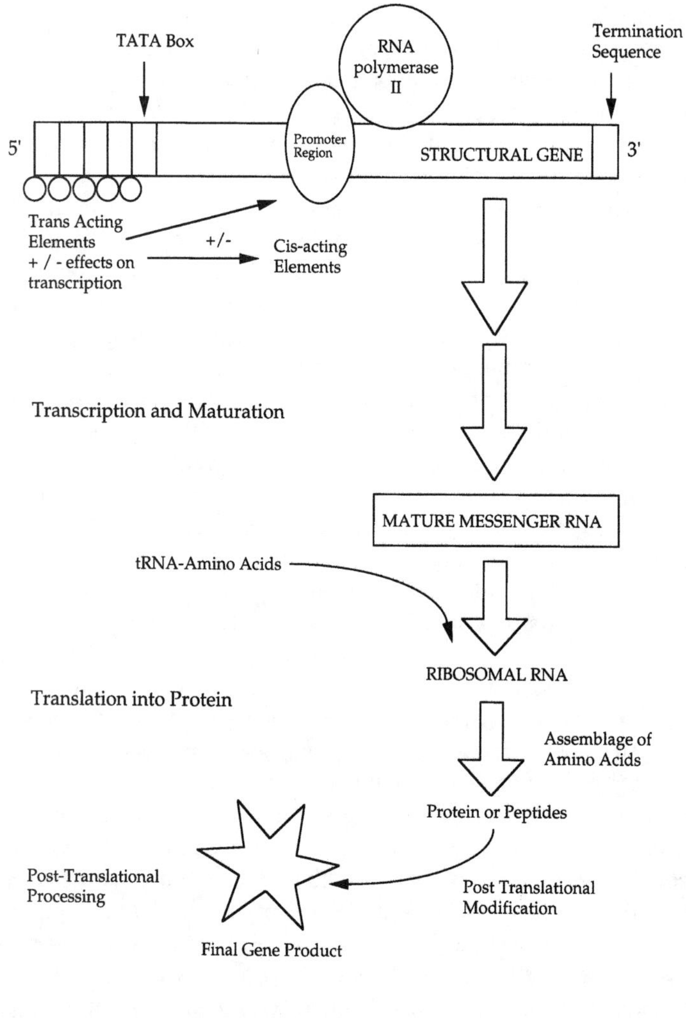

FIGURE 1. Gene expression is controlled at a number of places in the pathway from the nuclear DNA to the completed gene product. Transcription control is usually stronger than the control of translation. This control is exerted by the promoter region which in turn is influenced by cis and trans elements. Some promoters, i.e., TATA, GC, and CCAAT boxes are common to many different genes transcribed by RNA polymerase II. These elements, usually proteins, can have negative (inhibitory) or positive (stimulatory) effects on gene transcription. Such proteins labeled jun, AP2, ATF, CREB, SPI, OTFI, CTF, NFI, and SRF have been identified as enhancer elements. These elements can be glycoproteins with a specific carbohydrate as part of its structure or lipoproteins with a specific lipid such as a steroid hormone or a vitamin or a fatty acid as part of its structure. Small peptides, polypeptides, or simple proteins having an affinity for specific base sequences in the DNA can serve also in this capacity. Translation is controlled as well. The availability of the appropriate amounts of amino acids, the stability of the mRNA, and the sequence of the mRNA can affect translation efficiency.

with a series of zinc fingers. As mentioned, it recognizes a specific DNA sequence. For some members of this family of elements, the transcription-enhancing domain is localized at the amino terminus of the polypeptide chain. At the carboxy terminus is the binding site for the steroid hormone.

Post-transcriptional regulation of gene expression is the next stage of control for many genes. RNA transcription can be terminated prematurely with the result of a smaller than expected gene product. The mRNA can also be spliced to produce different proteins from the same gene. These proteins may have comparable functions. Actually, most RNAs are far larger than the mature RNA generated from it by splicing and only 1/20th of this RNA ever leaves the nucleus. The remainder is degraded and the nucleosides are reused or subjected to further degradation. The RNA that leaves the nucleus does so through pores in the nuclear membrane. This is an active process of which the details are not well known.

Not all of the mRNA that exits the nucleus is translated into protein. Translation can be blocked by specific proteins that bind at sites near the 5′ end of the molecule. This binding exerts negative translational control on gene expression. The mRNA has been made *but* the protein is not made. An example of this is seen in the regulation of the synthesis of ferritin. Ferritin mRNA is not translated when an iron response element binds to it. This allows for a rapid shift in ferritin synthesis when iron is present and an equally rapid shift away from ferritin synthesis when iron is in short supply. When iron is present, the iron response element folds away from the start site for translation making it available for use. When iron is absent, this start site is covered up by the iron response element which serves as a negative control element. The details of this process were described in Volume 1. Many mRNAs are subject to translational control by nutrients in this fashion.

The mRNAs have a very short half life when compared to DNA and the other RNAs. If the half life is shortened or prolonged, gene expression is affected. Many of the very unstable mRNAs have half lives in terms of minutes — among these are those which code for short-lived regulatory proteins such as the protooncogenes, fos and myc. This instability is probably due to an A and U-rich 3′ untranslated region. Stability can be affected by steroid hormones, nutritional state, and drugs.

Once the mRNA has migrated from the nucleus to the cytoplasm and attaches to ribosomes, translation is ready to begin. All of the amino acids needed for the protein being synthesized must be attached to a transfer RNA (tRNA). These tRNA amino acids dock on the mRNA again using base pairing and the amino acids are joined to one another via the peptide bond. The newly synthesized protein is released as it is made on the ribosomes, and changes to its conformation and structure occur. These changes depend on the constituent amino acids and their sequence. Post-translational modification includes a wide variety of change. For example, nuclear encoded proteins needed to complete OXPHOS are synthesized with a leader that allows them to migrate into the mitochondria. This leader is then removed as the OXPHOS system is assembled. Another example is prothrombin which is assembled with a large number of glutamic acid residues. In the presence of vitamin K these residues are carboxylated and this post-translational change results in a dramatic increase in the calcium binding capacity of the resultant protein. Unless prothrombin can bind calcium it cannot function in the clotting process. This is another example of how a nutrient can affect gene expression, in this instance, the expression of functional prothrombin. The site of the nutritional effect is that of post-translational protein modification. Extending this further, one could argue that nutrients in surplus or excess also affect gene expression by altering the milieu in which gene products must function. For example, dietary fat intake can influence membrane composition, which in turn can affect the activity, and indeed, the conformational structure of the proteins embedded in these membranes.[50]

Lastly, nutrition can affect the phenotypic expression of the genotype if certain of these nutrients are in short supply. This is the basis for the dietary management of a number of genetically determined disorders. For example, dietary phenylalanine is limited in the management of phenylketonuria to prevent the typical phenotypic expression of the defective genotype, mental retardation. Perusal of the disorders listed in Table 1 will suggest other conditions that can be managed in this way: lactose/lactase deficiency; galactose/galactosemia; carbohydrate/diabetes mellitus and so forth.

Knowledge about some of the single gene defects has contributed to and has been expanded by the scientific advances that are part of the human genome project. It is the purpose of this project to completely sequence and map the human genome. In the following section, one disease, maple syrup urine disease, is described so that the reader can appreciate the variation of etiology of this single disease syndrome.

MAPLE SYRUP URINE DISEASE

The above general discussion of the control of gene expression is a necessary prologue to the in-depth discussion of a disorder called maple syrup urine disease. Five phenotypes have been described based on the severity of symptoms, dietary protein tolerance, and enzyme activity. In its most severe form, people with this autosomal recessive disorder in the degradation of the branched-chain amino acids (BCAA) are characterized by a rapid onset of oxoacidosis within the first week of life. This is accompanied by urine with the odor of maple syrup and with seizures, coma, and death if untreated. Milder forms of the disorder result in mental retardation and some patients respond to thiamin supplementation. The disorder is due to one or more mutations in the genes which encode the various components of the branched-chain α keto acid dehydrogenase complex (BCKDH). Once the branched-chain amino acids (leucine, isoleucine, and valine) are deaminated, they are then oxidatively decarboxylated through the activity of this enzyme complex. This is located in the mitochondrial compartment and consists of three catalytic components: a branched-chain α keto acid decarboxylase called A1, which is a heterotetramer composed of two E1α and two E1β subunits. This component catalyzes the decarboxylation of the branched-chain α keto acids. The second component is called E2 and is a dihydrolipid transacylase. This component comprises the structural core of the enzyme complex to which the other components are attached by noncovalent interactions. The third component, E3, is a dihydrolipoyl dehydrogenase, a flavoprotein that is shared with pyruvate dehydrogenase and α ketoglutarate dehydrogenase. The complex also contains two regulatory enzymes: a specific kinase and a specific phosphatase that regulate complex activity through a phosphorylation–dephosphorylation cycle. Chapter 2 discusses this regulation.

The genes which encode all of these components have been identified, cloned, and partly mapped.[51-54] The E1β gene has been mapped to chromosome 6p21-22 while the E2 gene has been mapped to chromosome 1p31. The E2 gene has been sequenced.[55] The coordinate regulation of transcription of these genes is unknown. However, there is striking amino acid homology between the subunits of this complex and the subunits of the pyruvate dehydrogenase (PDH) complex. Zneimer et al.[53] have suggested that each of the BCKDH and PDH subunit loci belong to a distinct multigene family that arose from duplication and subsequent diversification of ancestral genes. The chromosome map (Chromosome 1) where the E2 gene resides is a well-studied region. Assigned to this region are genes for the major histocompatibility complex (MHC), tumor necrosis factor (TNF), a heat shock protein (HSP 70), prolactin, collagen-type X1α2, cytochrome P450, steroid 21-hydroxylase, hemachromatosis complement components 2, 4A, and 4B, properdin factor B, glyoxalase 1, motilin, pim oncogene homolog, methylmalonyl coenzyme A mutase, tRNA methionine initiator, and microtubule β-associated γ-like protein. One of the glucose transporters (GLUT 5) has also been mapped to this region.

Genetic heterogeneity is common in patients with maple syrup urine disease and probably accounts for the differences in severity of the symptoms. Several different mutations have been identified. In one patient, a 17 bp insertion in the E2 gene due to aberrant splicing has been described. This mutation results in a missense codon (T→G) that changes Phe 215 to Cys in the E2 subunit.[56] This mutation is thiamin-responsive in that doses of 10 mg of thiamin per day normalized the blood levels of the branched chain amino acids. Another mutation consists of a Tyr-393 to Asn substitution in the E1α and β subunits which in turn leads to their degradation. This mutation was found in 1 of 176 live births in a Mennonite population in Philadelphia. This particular mutation has been studied in detail by Chuang and associates and further, they devised a means to correct it.[58,59] Using a retroviral gene transfer technique, they were able to correct the E1α subunit mutation in lymphoblasts. Following this procedure, normal levels of branched-chain amino acids were observed in the blood, indicating that decarboxylation was proceeding normally. The patient was followed for more than three months and Southern dot blot analysis showed that the insert was fully integrated into the host genome. Subsequent Western and Northern blotting showed that both the normalized and mutant genes were expressing themselves. However, the presence of normal E1α subunit seemed to provide sufficient activity to prevent the symptoms of the disease.

Other mutations in BCKDH include one in the regulatory segment of E1α mRNA.[58] This mutation results in a Tyr-368 to Cys in the E1α subunit. Another results in a frame shift at codon 4 of the mature E1α subunit with the generation of a premature stop codon (TGA) at base 174 of the E1α mRNA.[59] Altogether, these mutations either decrease the steady state mRNA or impair E1 assembly and thus result in inadequate E1 function.

E2 mutations likewise result in impaired BCKDH function. A 2 bp deletion at exon 2 and a nonsense (G→T) transition in the E2 gene at exon 6 have been reported.[60] These mutations have as their result an array of single and multiple exon deletions. The G to T transition produces a premature stop codon at Glu 163 while the deletion at exon 2 causes a frameshift downstream of residue (–26) in the mitochondrial targeting presequence.[60,61] This, of course, affects the import of this subunit into the mitochondria and its assemblage into the BCKDH complex. As will be discussed in Chapter 2, all the subunits must be present in order to have full complex activity. This activity is carefully controlled so that degradation of the branched-chain amino acids is active only to the extent that these nutrients are in surplus. However, as indicated above, should a mutation occur in any of the genes which encode the subunits of this complex, the symptoms of maple syrup urine disease will appear.

SUMMARY

Aberrant products of gene expression account for a wide variety of human disease. Depending upon the location of the gene product, the results of such expression can be severe, moderate, or mild. The aberration in the gene can even be undetected until several decades of human life have passed. Vital metabolic pathways may have fail-safe compensatory mechanisms that sidestep or bypass a critical reaction in that pathway. When such compensatory mechanisms are not available and the mutation is such that the pathway cannot function, the viability of the cell or organism or individual is in jeopardy. Such severe gene defects probably explain early embryonic death. Fetal death, neonatal death, and death in infancy or childhood due to genetic errors likewise may be explained by mutations in genes which encode proteins (enzymes, carriers, receptors, etc.) that play vital roles in metabolism and metabolic control. Genetic errors or mutations in the genetic material, DNA, that manifest themselves later in life — adolescence and adulthood — are the ones that concern the nutrition scientist. For many of these, the composition and quantity of the food that is consumed can modify the phenotypic expression of the genotype. Extending this concept to the normal population that distributes along a bell-shaped curve with respect to any measurement, be it height, weight, disease status, or cause of death, one could argue that the ideal nutrient intake for any *individual* is dependent on that individual's genetic heritage. Ideally, one should design a diet which potentiates the expression of the individual's genes for normal growth, development, and longevity while suppressing the expression of genes that are associated with disease. This ideal is far from reality given the present state of knowledge about the effects of nutrients on gene expression. Nonetheless, such an approach is not an unused one when the management of single gene defects is considered. In the future, as we gain greater understanding of our genetic heritage and how specific nutrients can be used to manipulate the results of that heritage, we should see an overall improvement in healthy lifespans.

REFERENCES

1. **Berdanier, C. D. and Hargrove, J. L.,** *Nutrition and Gene Expression,* CRC Press, Boca Raton, 1993.
2. **Shoffner, J. M. and Wallace, D. C.,** Oxidative phosphorylation disease and mitochondrial DNA mutations: diagnosis and treatment, *Ann. Rev. Nutr.,* 14, 535, 1994.
3. **Anderson, S., Bankier, A. T., Barrell, B. G., Bruijin, M. H. L., Coulsen, A. R., Drouin, J., Eperon, I. C., Nierlich, D. P., Roe, B. A., Sanger, F., Schreier, P. H., Smith, A. J. H., Staden, R., and Young, I. G.,** Sequence and organization of the human mitochondrial genome, *Nature,* 290, 456, 1981.
4. **Cantatore, P. and Sttardi, G.,** Mapping the nascent light and heavy strand transcripts on the physical map of HeLa cell mtDNA, *Nucl. Acid Res.,* 8, 2605, 1980.
5. **Chang, D. D. and Clayton, D. A.,** Precise identification of individual promoters for transcription of each strand of human mtDNA, *Cell,* 36, 635, 1984.
6. **Monnat, R. J. and Reay, D. T.,** Nucleotide sequence identity of mitochondrial DNA from different human tissues, *Gene,* 43, 205, 1986.
7. **Hixson, J. E. and Clayton, A.,** Initiation of transcription from each of the two human mitochondrial promotes requires unique nucleotide as transcriptional sites, *Proc. Natl. Acad. Sci. U.S.A.,* 82, 2660, 1985.
8. **Foran, D. R., Hixson, J. E., and Brown, W. M.,** Comparisons of ape and human sequences that regulate mtDNA transcription and D-loop DNA synthesis, *Nucl. Acid Res.,* 16, 581, 1988.
9. **Giles, R. E., Blanc, H., Cann, H. M., and Wallace, D. C.,** Maternal inheritance of human mtDNA, *Proc. Natl. Acad. Sci. U.S.A.,* 77, 6715, 1980.
10. **Shoubridge, E. A.,** Mitochondrial DNA diseases: histological and cellular studies, *J. Bioenergetics Biomembranes,* 26, 301, 1994.
11. **Madsen, C. S., Ghivizzani, S. C., and Hauswirth, W. W.,** *In vivo* and *in vitro* evidence for slipped mispairing in mammalian mitochondria, *Proc. Natl. Acad. Sci. U.S.A.,* 90, 7671, 1993.
12. **Degoul, F., Nelson, I., Amselem, S., Romero, N., Obermaier-Kusser, B., Ponsot, G., Marsac, C., and Lestienne, P.,** Different mechanisms inferred from sequences of human mitochondrial DNA deletions in ocular myopathies, *Nucl. Acids Res.,* 19, 493, 1990.

13. **Zeviana, M., Bresolin, N., Gellera, C., Bordoni, A., Pannacci, M., Amati, P., Oggio, M., Servidei, S., Scarlato, G., and DiDonato, S.,** Nucleus driven multiple large scale deletions of the human mitochondrial genome: a new autosomal dominant disease, *Am. J. Hum. Genet.,* 47, 904, 1990.
14. **Lestienne, P.,** Mitochondrial DNA mutations in human disease: a review, *Biochimie,* 74, 123, 1992.
15. **Kapsa, R., Thompson, G. N., Thorburn, D. R., Dahl, H. H. M., Marzuki, S., Byrne, E., and Blok, R. B.,** A novel mtDNA deletion in an infant with Pearson Syndrome, *J. Inher. Metab. Dis.,* 17, 521, 1994.
16. **Yen, T-C., Su, H-H., King, K-L., and Wei, Y-H.,** Aging associated 5 kb deletion in human liver mitochondrial DNA, *Biochem. Biophys. Res. Commun.,* 178, 124, 1991.
17. **Yen, T-C., Pang, C-Y., Hsieh, R-H., Su, C-H., King, K-L., and Wei, Y-H.,** Age dependent 6 kb deletion in human mitochondrial DNA, *Biochem. Int.,* 26, 457, 1992.
18. **Corral-Debrinski, M., Stepien, G., Shoffner, J. M., Lott, M. T., Kanter, K., and Wallace, D. C.,** Hypoxemia is associated with mtDNA damage and gene induction. Implications for cardiac disease, *JAMA,* 266, 1812, 1991.
19. **Simonetti, S., Chen, X., DiMauro, S., and Schon, E. A.,** Accumulation of deletions in human mitochondria during normal aging: analysis by quantitative PCR, *Biochem. Biophys. Acta,* 1180, 113, 1992.
20. **Yen, T-C., Chen, Y-S., King, K-L., Yeh, S-H., and Wei, H-H.,** Liver mitochondrial respiratory function declines with age, *Biochem. Biophys. Res. Commun.,* 165, 994, 1989.
21. **Asano, K., Nakamara, M., Sato, T., Tauchi, H., and Asano, A.,** Age dependency of mtDNA decrease differs in different tissues of rat, *J. Biochem.,* 114, 303, 1993.
22. **Calleja, M., Pena, P., Ugalde, C., Fereiro, C., Marco, R., and Garesse, R.,** Mitochondrial DNA remains intact during Drosophila aging, but the levels of mitochondrial transcripts are significantly reduced, *J. Biol. Chem.,* 268, 18891, 1993.
23. **Silvestri, G., Moraes, C. T., Shanske, S., Oh, S. J., and DiMauro, S.,** A new mtDNA mutation in the tRNA lys gene associated with myoclonic epilepsy and ragged red fibers (MERRF), *Am. J. Hum. Genet.,* 51, 1213, 1992.
24. **Reardon, W., Ross, R., Sweeney, M. G., Luxon, L. M., Pembrey, M. E., Harding, EA. E., and Trembath, R. C.,** Diabetes mellitus associated with a pathogenic point mutation in mtDNA, *Lancet,* 340, 1376, 1992.
25. **Gerbitz, K. D.,** Does the mitochondrial DNA play a role in the pathogenesis of diabetes? *Diabetologia,* 35, 1181, 1992.
26. **Gerbitz, K. D., Paprotta, A., Jaksch, M., Zierz, S., and Drechsel, J.,** Diabetes mellitus is one of the heterogeneous phenotypic features of a mtDNA point mutation within the tRNA leu gene, *FEBS Lett.,* 321, 194, 1993.
27. **Wakeley, J.,** Substitution rate variation among sites in hypervariable region 1 of human mtDNA, *J. Mol. Evol.,* 37, 613, 1993.
28. **Howell, N., McCullough, D. A., Kubacka, I., Halvorson, S., and Mackey, D.,** The sequence of human mtDNA: the question of errors versus polymorphisms, *Am. J. Hum. Genet.,* 50, 1333, 1992.
29. **Tanaka, M. Miyabayashi, S., Nishikimi, M., Suzuki, H., Shixomura, Y., Ito, K., Narisawa, K., Tada, K., and Ozawa, T.,** Extensive defects in mitochondrial electron-transfer chain in muscular cytochrome c oxidase deficiency, *Ped. Res.,* 24, 47, 1988.
30. **Ewart, G., Lightowlers, R., Zhang, Y-Z., Balan, V. J., Kennaway, N., and Capaldi, R. A.,** Tissue specificity and defects in human cytochrome c oxidase, *Biochem. Biophys. Acta,* 1018, 223, 1990.
31. **DiMauro, S., Lombes, A., Nakase, H., Mita, S., Fabrizi, A. F., DeVivo, D. C., and Schon, E. A.,** Cytochrome c oxidase deficiency, *Ped. Res.,* 28, 536, 1990.

32. **Kennaway, N. G., Carrero-Valenzuela, R. D., Ewart, G., Balan, V. K., Lightowlers, R., Zhang, Y. Z., Powel, B. R., Capaldi, R. A., and Buist, N. R. M.,** Isoforms of mammalian cytochrome c oxidase: correlation with human cytochrome c oxidase deficiency, *Ped. Res.,* 28, 529, 1990.
33. **Moraes, C. T., Rici, E., Bonilla, E., DiMauro, S., and Schon, E. A.,** The mitochondrial tRNA leu (UUR) mutation in mitochondrial encephalomyopathy, lactic acidosis, and stroke like episodes (MELAS): genetic, biochemical, and morphological correlations in skeletal muscle, *Am. J. Hum. Genet.,* 50, 934, 1992.
34. **Lertrit, P., Noer, A. S., Jean-Francois, M. J. B., Kapsa, R., Dennett, X., Thyagarajan, D., Lethlean, K., Byrne, E., and Marzuki, S.,** A new disease-related mutation for mitochondrial encephalopathy, lactic acidosis and stroke like episodes (MELAS) syndrome affects the ND4 subunit of the respiratory complex 1, *Am. J. Hum. Genet.,* 51, 457, 1992.
35. **Inui, K., Fukushima, H., Tsukamoto, H., Taniike, M., Midoriekawa, M., Tanaka, J., Nishigaki, T., and Okada, S.,** Mitochondrial encephalopathies with the mutation of the mitochondrial tRNA leu (UUR) gene, *J. Pediatr.,* 120, 62, 1992.
36. **Moraes, C. T., Shanske, S., Tritschler, H-J., Aprille, J., Andreetta, F., Bonilla, E., Schon, E. A., and DiMauro, S.,** mtDNA depletion with variable tissue expression: a novel genetic abnormality in mitochondrial diseases, *Am. J. Hum. Genet.,* 48, 492, 1991.
37. **Wyburg, F. A., Wanders, R. J. A., van Lie Peters, E. M., Vos, G. D., Loggers, H. G., Bolhuis, P. A., Herzberg, N. H., Ruitenbeek, W., van Wilsem, A., ten Houten, R., and Barth, P. G.,** NADH:Q1 oxidoreductase deficiency without lactic acidosis in a patient with Leigh syndrome: implications for the diagnosis of unborn errors of the respiratory chain, *J. Inherit. Metab. Dis.,* 14, 297, 1991.
38. **Wallace, D. C., Zheng, X., Lott, M. T., Shoffner, J. M., Hodge, J. A., Kelley, R. I., Epstein, C. M., and Hopkins, L. C.,** Familial mitochondrial encephalomyopathy (MERRF): genetic pathophysiological and biochemical characterization of a mitochondrial DNA disease, *Cell,* 55, 601, 1988.
39. **Wallace, D. C., Lott, M. T., Lezza, A. M. S., Seibel, P., Voljavec, A. S., and Shoffner, J. M.,** Mitochondrial DNA mutations associated with neuromucular diseases: analysis and aging using the polymerase chain reaction, *Ped. Res.,* 28, 525, 1990.
40. **Slipetz, D. M., Goodyer, P. R., and Rozen, R.,** Congenital deficiency of a 20 kDa subunit of mitochondrial complex I in fibroblasts, *Am. J. Hum. Genet.,* 48, 1121, 1991.
41. **Tulinius, M. H., Holme, E., Kristiansson, B., Larsson, N-G., and Oldfors, A.,** Mitochondrial encephalomyopathies in childhood, I: biochemical and morphologic investigations; II: clinical manifestations and syndromes, *J. Pediatr.,* 119, 242, 1991.
42. **Scholte, H. R., Agsteribbe, E., Busch, H. F. M., Hoogenraad, T. U., Jennekens, F. G. I., vanLinge, B., Luyt-Houwen, I. E. M., Ross, J. D., Ruiters, M. H. J., and Verduin, M. H. M.,** Oxidative phosphorylation in human muscle in patients with ocular myopathy and after general anesthesia, *Biochem. Biophys. Acta,* 1018, 211, 1990.
43. **Savontaus, M-L., Huoponen, K., Majander, A., Aula, P., and Nikoskelainen, E. K.,** Gene defects in Leber's hereditary optic neuroretinopathy, *Biochem. Biophys. Acta,* 1101, 204, 1992.
44. **Brown, M. D., Voljavec, A., Lott, M. T., MacDonald, I., and Wallace, D. C.,** Leber's hereditary optic neuropathy: a model for mitochondrial neurodegenerative diseases, *FASEB J.,* 6, 2791, 1992.
45. **Huoponene, K., Lamminen, T., Juvonen, V., Aula, P., Nikoskelainen, E., and Savontaus, M-L.,** The spectrum of mtDNA mutations in families with Leber's hereditary optic neuropathy, *Hum. Genet.,* 92, 379, 1993.
46. **Derman, E.,** Transcriptional control in the production of liver specific mRNA's, *Cell,* 23, 731, 1981.
47. **Maniatis, T., Goodbourn, S., and Fischer, J. A.,** Regulation of inducible and tissue specific gene expression, *Science,* 236, 1237, 1987.

48. **Atchison, M. L.,** Enhancers: mechanisms of action and cell specificity, *Ann. Rev. Cell. Biol.,* 4, 127, 1988.
49. **Yamamoto, K.,** Steroid receptor regulated transcription of specific genes and gene networks, *Ann. Rev. Genet.,* 19, 209, 1985.
50. **Berdanier, C. D.,** Faty acids and membrane function, in *Fatty Acids in Foods and Their Health Implications,* Chow, C. K., Ed., Marcel Dekker, New York, 1992, 531.
51. **Lau, K. S., Eddy, R. L., Shows, T. B., Fisher, C. W., Chuang, D. T., and Cox, R. P.,** Localization of the dihydrolipoamide branched chain transacylase gene (DBT) of the human branched-chain keto acid dehydrogenase complex to chromosome 1, *Cytogenet. Cell. Genet.,* 56, 33, 1991.
52. **Chuang, D. T., Fisher, C. W., Lau, K. S., Griffin, T. A., Wynn, R. M., and Cox, R. P.,** Maple syrup urine disease: domain structure, mutations and exon skipping in the dihydrolipoyl transacylase (E2) component of the branched chain α keto acid dehydrogenase complex, *Mol. Biol. Med.,* 8, 49, 1991.
53. **Zneimer, S. M., Lau, K. S., Eddy, R. L., Shows, T. B., Chuang, J., Chuang, D. T., and Cox, R. P.,** Regional assignment of two genes of the human branched chain α keto acid dehydrogenase complex: the E1β gene (BCKDHB) to chromosome 6p21-22 and the E2 gene (DBT) to chromosome 1p31, *Genomics,* 10, 740, 1991.
54. **Chuang, D. T.,** Molecular studies of mammalian branched chain α keto acid dehydrogenase complexes: domain structures, expression, and inborn errors, *Ann. N.Y. Acad. Sci.,* 573, 137, 1989.
55. **Lau, K. S., Chuang, J. L., Herring, W. J., Danner, D. J., Cox, R. P., and Chuang, D. T.,** The complete cDNA sequence for dihydrolipoyl transacylase (E2) of human branched-chain α-keto acid dehydrogenase complex, *B. B. A.,* 1132, 319, 1992.
56. **Fisher, C. W., Lau, K. S., Fisher, C. R., Wynn, R. M., Cox, R. P., and Chuang, D. T.,** A 17 bp insertion and a Phe 215→Cys missense mutation in the dihydrolipoyl transacylase (E2) mRNA from a thiamine-responsive maple syrup urine disease patient WG-34, *B.B.R.C.,* 174, 804, 1991.
57. **Fisher, C. R., Chuang, J. L., Cox, R. P. Fisher, C. W., Star, R. A., and Chuang, D. T.,** Maple syrup urine disease in Mennonites: evidence that the Y393N mutation in E1α impedes assembly of the E1 component of branched chain α keto acid dehydrogenase complex, *J. Clin. Invest.,* 88, 1034, 1991.
58. **Chuang, J. L., Fisher, C. R., Cox, R. P., and Chuang, D. T.,** Molecular basis of maple syrup urine disease: novel mutations at the E1α locus that impair E1($\alpha_2\beta_2$) assembly or decrease steady state E1α mRNA levels of branched chain α keto acid dehydrogenase complex, *Am. J. Hum. Genet.,* 55, 297, 1994.
59. **Koyata, H., Cox, R. P., and Chuang, D. T.,** Stable correction of maple syrup urine disease in cells from a Mennonite patient by retroviral-mediated gene transfer, *Biochem. J.,* 295, 635, 1993.
60. **Fisher, C. W., Fisher, C. R., Chuang, J. L., Lau, K. S., Chuang, D. T., and Cox, R. P.,** Occurrence of a 2-bp (AT) deletion allele and a nonsense (G→T) mutant allele at the E2 (DBT) locus of six patients as a secondary effect of mutations, *Am. J. Hum. Genet.,* 52, 414, 1993.
61. **Lau, K. S., Lee, J., Fisher, C. W., Cox, R. P., and Chuang, D. T.,** Premature termination of transcription and alternative splicing in the human transacylase (E2) gene of the branched chain α keto acid dehydrogenase complex, *FEBS Lett.,* 279, 229, 1991.

Chapter 2

REGULATION OF GENE EXPRESSION OF THE BRANCHED-CHAIN KETO ACID DEHYDROGENASE COMPLEX

Harbhajan S. Paul

CONTENTS

INTRODUCTION

The mammalian branched-chain keto acid dehydrogenase (BCKDH) complex catalyzes the oxidative decarboxylation of the branched-chain keto acids (BCKA) α-ketoisocaproate, α-ketoisovalerate, and α-keto-β-methylvalerate which are derived from transamination of the branched-chain amino acids (BCAA) leucine, valine, and isoleucine, respectively. BCKDH is a mitochondrial multienzyme complex composed of three catalytic proteins. These include BCKA decarboxylase (E1), dihydrolipoyl transacylase (E2), and dihydrolipoyl dehydrogenase (E3).[1-3] The E1 component is further composed

0-8493-9485-6/96/$0.00+$.50

of two α and two β subunits.[4] The E1 and E2 components are specific for BCKDH, whereas E3 is common to BCKDH, pyruvate dehydrogenase, and α-ketoglutarate dehydrogenase complexes.[5] The BCKDH complex is also associated with two specific regulatory proteins, a kinase and a phosphatase that modulate the activity of the complex through a reversible phosphorylation (inactivation) and dephosphorylation (activation) cycle.[6-8] The proportion of the enzyme complex that exists in active dephosphorylated form (activity state of BCKDH) varies in a tissue-specific manner.

BCKDH subunits are encoded by nuclear genes and synthesized on cytoplasmic ribosomes. Each subunit has a leader peptide which allows these proteins to be imported into mitochondria. Once inside the mitochondria, the leader peptide is cleaved and the subunits assemble to form a functional BCKDH complex.[1,3] Evidence is emerging from recent studies that molecular chaperones may be involved in correct folding and assembly of subunits of BCKDH complex.[9]

BCAA have a critical role in nutrition and play an important role in metabolic regulation. Studies have shown that BCAA may regulate a wide variety of metabolic processes, such as protein synthesis, insulin secretion, transport of amino acids across blood–brain barrier, etc.[10] Therefore, advances in the understanding of regulation of BCAA metabolism may have application to the understanding of the pathogenesis of common clinical and metabolic conditions. There is also considerable biochemical and clinical interest in the regulation of this enzyme at both the biochemical and molecular levels because of reduced activity of this enzyme complex in patients with maple syrup urine disease.[1]

During the past two decades, much has been learned about the metabolism of BCAA in mammalian tissues.[1,7,8] These studies have revealed that BCKDH is the rate-limiting enzyme which commits BCAA to their catabolic fate. The activity state of the enzyme varies among tissues. For example, BCKDH exists mostly in the dephosphorylated active form in the liver, whereas it exists largely in the phosphorylated inactive form in skeletal muscle.[11,12] In recent years, studies have resulted in the purification and characterization of BCKDH and production of specific antibodies.[6,13,14] The application of recombinant DNA technology has resulted in the cloning of cDNAs for various subunits of rat and human BCKDH.[1,3] The availability of these tools has now made it possible to investigate the regulation of the BCKDH complex at the molecular level.

REGULATION OF THE BCKDH COMPLEX

At the present time, three mechanisms are known to regulate the activity of the mammalian BCKDH complex. These include modulation of enzyme activity by substrate analog and products, reversible phosphorylation–dephosphorylation, and changes in gene expression.

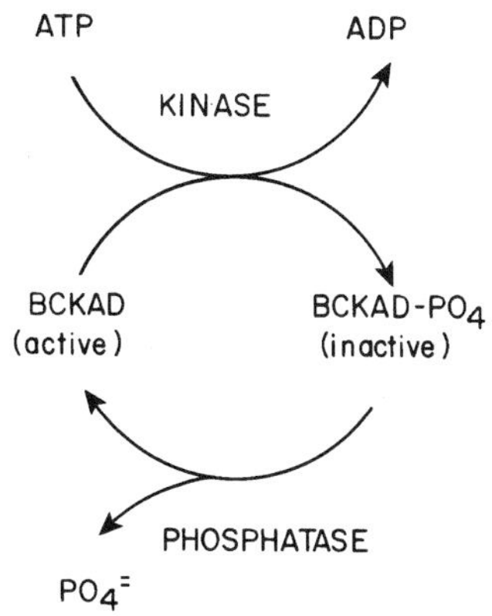

FIGURE 1. Regulation of BCKDH by phosphorylation and dephosphorylation cycle.

MODULATION BY-PRODUCTS AND SUBSTRATE ANALOG

The BCKDH complex is subject to feedback inhibition by NADH and branched-chain acyl-CoA.[1,8] Substrates such as fatty acids and pyruvate, by altering the NADH:NAD$^+$ ratio, affect the flux through BCKDH complex.[15,16] Carnitine stimulates the oxidation of BCAA by removing the branched-chain acyl-CoA as branched-chain acylcarnitine.[17,18] Octanoate and hypolipidemic agents such as clofibrate increase oxidation of BCAA by inhibiting the activity of BCKDH kinase.[12,19,20]

PHOSPHORYLATION–DEPHOSPHORYLATION

Several laboratories have established that the activity of the BCKDH complex is regulated by reversible phosphorylation–dephosphorylation cycle (Figure 1). As pointed out above, a specific protein kinase catalyzes phosphorylation of specific serine residues on E1α subunit and inactivates the enzyme complex.[21] Conversely, dephosphorylation of BCKDH catalyzed by a specific phosphoprotein phosphatase activates this enzyme.[22] By using a broad-specificity phosphoprotein phosphatase to dephosphorylate the BCKDH complex, total activity of BCKDH can be determined. Studies have shown that nearly 100% of the enzyme complex in the liver is in the active (dephosphorylated) form whereas in skeletal muscle only 1 to 2% of the enzyme complex is in the active form.[11,12] The activity state of the enzyme in other tissues such as kidney and heart ranges between the above values.[11] Differences in activity state among tissues appear to be related to the levels and activity of the BCKDH kinase.[23]

Nutritional and hormonal alterations play an important role in modulating the activity of BCKDH by a phosphorylation–dephosphorylation cycle. For example, rats fed a low protein diet show low hepatic BCKDH activity.[24,26] This decrease in complex activity is accompanied by a reduction in the activity state of the enzyme complex compared to rats fed a diet containing adequate protein. Miller et al.[27] have shown that liver and muscle mitochondria from rats fed a high protein diet contain largely the dephosphorylated form of E1α subunit of the BCKDH complex. In contrast, mitochondria from rats fed a low protein diet contain predominantly a phosphorylated form of the enzyme complex.[27] The effect of dietary protein on the activity state of BCKDH is mediated by the intracellular levels of BCKA, which affect the activity of BCKDH kinase.[21] A high protein diet increases the tissue levels of BCKA which then inhibit the BCKDH kinase activity. Inhibition of the kinase results in the complex being in the dephosphorylated active form. This situation allows for the rapid disposal of excess BCAA. In contrast, in animals fed a low protein diet, the tissue levels of BCKA decrease such that the kinase activity is no longer suceptible to inhibition. In this situation, the complex is converted into a phosphorylated inactive form. Phosphorylation and inactivation of the BCKDH complex in response to a low protein diet helps conserve BCAA for protein synthesis in this nutritional state. Also, alteration of the phosphorylation state permits rapid modulation of enzyme activity.

The above studies led to the thinking that post-translational modification of BCKDH by phosphorylation and dephosphorylation is the basis for regulation of this enzyme and that there is no induction of BCKDH in response to metabolic perturbations. However, as reviewed below, recent studies demonstrate that, besides covalent modification, changes in gene expression also contribute to the regulation of BCKDH activity.

REGULATION OF BCKDH GENE EXPRESSION:

Gene Expression During Differentiation

The activity of the BCKDH complex has been reported to increase by 10- to 20-fold during hormone-mediated differentiation of 3T3-L1 preadipocytes to adipocytes.[28,29] The increase in BCKDH activity was accompanied by concomitant increases in the protein mass and the mRNA levels encoding each of the subunits of the BCKDH complex (E1, E2, and E3).[30,31] There appears to be a coordinated pattern of increase in all three BCKDH subunit mRNAs and protein mass associated with differentiation.[31] The induction of BCKDH in 3T3-L1 cells during adipocyte differentiation is consistent with the lipogenic role of this enzyme in providing acetyl-CoA and other carbon fragments for lipid synthesis.

Gene Expression During Development

Significant increases in BCKDH activity during neonatal development have been observed.[31-33] Subsequently, it was shown that the increase in enzyme activity is due to an increase in the mass of BCKDH subunits rather than to a change in the activity state of the enzyme.[33] Chinsky et al.[31] found significant increases in mRNA levels for all three subunits in murine liver, kidney, and heart between the ages of 1 day and 7 weeks. Increases in mRNA levels during development are consistent with the increased BCKDH activity and subunit mass observed during the first weeks of life.[33] An increase in all three BCKDH subunit mRNAs associated with growth and development suggests a coordinated pattern of expression of BCKDH subunit genes. The low activity of BCKDH complex, its subunit mass, and mRNA levels during the suckling period, relative to that in the adult animal, suggest an adaptation to conserve BCAA for protein synthesis in the newborn and suckling animals.

Gene Expression in the Liver

Of all the tissues, liver is most active in the catabolism of BCKA. Therefore, regulation of BCKDH gene expression in this tissue is of considerable interest. However, gene expression of BCKDH in liver has been investigated only recently.[26,31] A recent study examined the effect of dietary protein level on the expression of BCKDH genes in the intact liver.[26] Rats were fed for 2 weeks either with a protein-free diet, a 50% protein diet, or a commercial chow. Initially, the effect of dietary protein on BCKDH activity was determined. Hepatic BCKDH activity was significantly lower in rats fed the protein-free diet as compared to rats fed either the commercial chow or the high protein diet.[26] Western blot analysis revealed no significant difference in the amount of E2 subunit among the groups of rats fed three diets.[26] In contrast, the mass of E1α subunit was found to be lower in livers of rats fed the low protein diet and higher in rats fed the high protein diet than chow-fed rats.[26] The effect of the dietary protein level on the E1β subunit was not analyzed. The same study also investigated the effect on BCKDH mass of feeding the high protein diet to rats conditioned first to consume a protein-free diet. The amounts of E1α and E2 proteins were found to decrease after 14 days on the protein-free diet as compared to chow-fed rats.[26] Within 3 days of feeding the 50% protein diet, the hepatic BCKDH subunit protein returned to the values of the chow-fed rats.[26] The change in the protein mass of E1α subunit was greater than that of the E2 subunit when rats were fed either a protein-free diet or a high protein diet.[26]

In order to account for changes in the amount of BCKDH subunits in response to dietary protein, these authors measured the mRNA levels for BCKDH subunits. They found that dietary protein resulted in a divergent pattern of expression of the mRNAs. Feeding a 50% protein diet resulted in an increase in the level of E1β mRNA, a decrease in the level of E1α mRNA, and no change in the level of E2 mRNA.[26] They concluded that coordinated regulation of mRNAs encoding BCKDH subunits does not occur in response to dietary protein but that post-transcriptional mechanisms also regulate BCKDH subunit levels.

Another study examined the effect of a high protein diet on BCKDH subunit mRNAs in murine tissues.[31] The steady state levels of mRNAs for two of the subunits of BCKDH (E1β and E2) decreased by two- to fourfold in the livers of mice fed 0% protein diet compared to mice fed standard (23%) or high (50%) protein diets. In contrast, the levels of mRNA encoding the E1α subunit did not change in response to changes in dietary protein.[31] Dietary effects on mRNA levels were reversible within 48 hours when dietary protein contents were changed from 0 to 50% protein.[31] Therefore, the mechanisms producing the variations in BCKDH mRNAs in response to changes in dietary protein are likely to be different from those utilized during development to produce generalized increases in BCKDH. Changes in mRNA levels appeared to be tissue-specific since no significant changes in mRNA levels were noted in kidney in response to changes in dietary protein.[31] Although Chinsky et al.[31] did not study the effect of dietary protein on BCKDH activity or its subunit mass, they suggested, based on non-coordinated changes in the mRNA levels, that multiple regulatory mechanisms modulate BCKDH subunit expression. Although a mechanism for coordinate regulation of expression of all three subunits of BCKDH would seem advantageous, the observed changes in BCKDH subunit masses and mRNA levels argue against such a mechanism.

From the results of the above studies, it seems reasonable to suggest that alterations in dietary protein affect gene expression of certain subunits of BCKDH, but the relation to altered enzyme activity remains unclear. Marked increases in portal blood glucagon concentrations are observed in rats after high-protein feeding.[34] Since the action of glucagon is mediated via cAMP, it is likely that the observed increases in total hepatic BCKDH activity in response to a high protein diet reflect the action of cAMP as discussed below.

Gene Expression in Primary Hepatocyte Culture

As compared to intact liver, culturing hepatocytes in a defined, hormone-free medium often results in low basal gene expression for metabolic enzymes,[35,36] thereby improving the ability to detect and study induction of enzymes by individual hormones or combination of hormones. Therefore, primary hepatocyte cultures are highly suitable for studies of gene expression. Recent studies have investigated the effect of dexamethasone and CPT-cAMP (a cAMP analog) on gene expression of BCKDH.[37] These agents were chosen because of their known role in regulating the expression of genes encoding several enzymes involved in hepatic amino acid metabolism.[36,38] Studies of gene expression by hormones were approached initially by measuring the rate of flux through BCKDH using intact hepatocytes.[37] Measurement of the flux rate has the advantage that it allows for rapid screening to evaluate the effect of hormones and other agents on enzyme activity. Both dexamethasone and CPT-cAMP increased the rate of flux through BCKDH in a dose-dependent manner.[37] Maximum effect of dexamethasone was observed at a concentration of 1 μM and that of CPT-cAMP at a concentration of 50 μM.[37]

Since there are reports of synergistic or additive effects of dexamethasone and cAMP on metabolic enzymes,[36] the above studies also investigated the effect of a combination of these agents on the flux through BCKDH. A significantly greater effect of dexamethasone plus CPT-cAMP on the flux through BCKDH was found than with either agent alone (Figure 2).

Hormonally-mediated increases in the rate of flux reflected increases in both the basal and total activities of BCKDH (Table 1). The increase in basal activity (2.2-fold) was comparable to the increase in total activity (2.0-fold). The activity state of BCKDH in hepatocytes was unaffected by dexamethasone

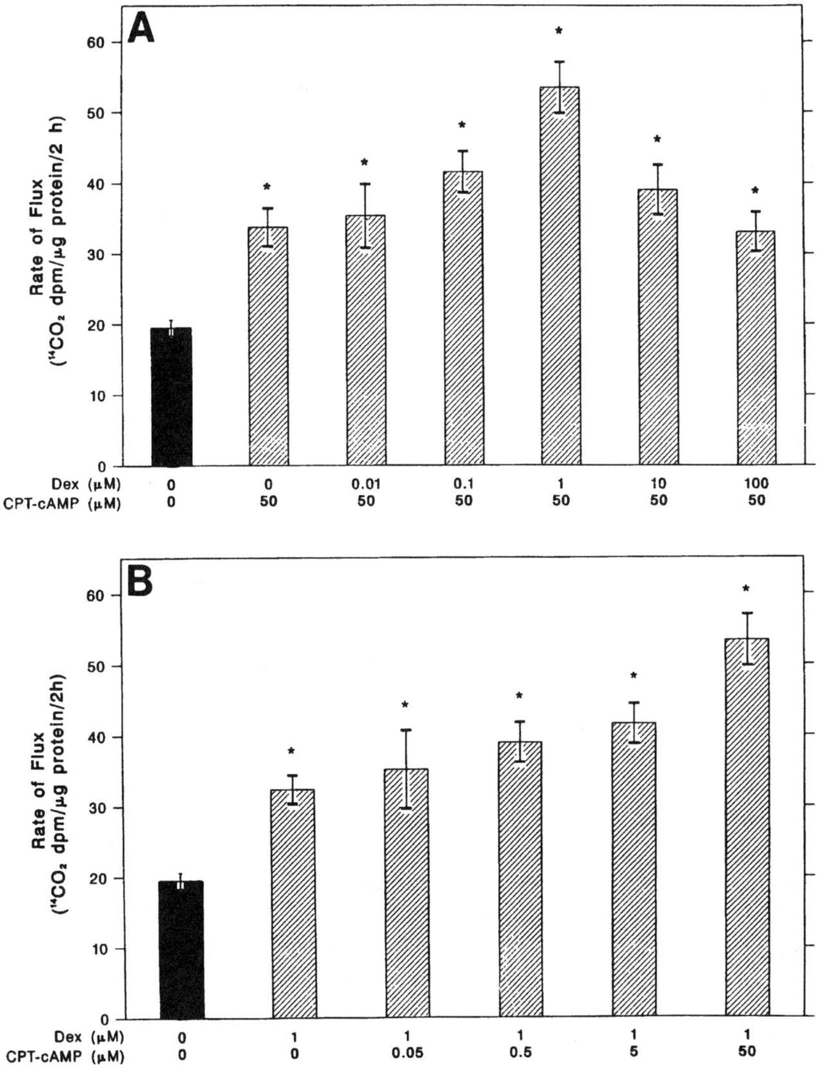

FIGURE 2. Effect of dexamethasone and CPT-cAMP combinations on the flux through BCKDH in cultured hepatocytes. Hepatocytes (0.5×10^5 cells/cm^2) were cultured for 24 h in the presence of the indicated combinations of dexamethasone and CPT-cAMP. Flux through BCKDH was determined by measuring the rate of $^{14}CO_2$ production from α-keto[1-^{14}C]isocaproate incubated with cultured hepatocytes for 2 h. The results are mean ± S.E. of hepatocyte cultures from 4 animals. **A.** The rate of flux was significantly higher (*; $P < 0.01$) in the presence of CPT-cAMP alone or in the presence of CPT-cAMP and all concentrations of dexamethasone studied than in control cultures. **B.** The rate of flux was significantly higher (*; $P < 0.01$) in the presence of dexamethasone alone or in the presence of dexamethasone and all concentrations of CPT-cAMP studied than in control cultures. (Reproduced from Chicco , A. G., Adibi, S. A., Liu, W. Q., Morris, S. M., and Paul, H. S., *J. Biol. Chem.*, 269, 19427, 1994. With permission.)

TABLE 1
Effect of Dexamethsone Plus CPT-cAMP on Basal and Total Activities of BCKDH and BCKDH Activity State in Primary Hepatocyte Culture

Treatment	BCKDH activity nmol/min/g wet weight		Activity State %
	Basal	Total	
Control	168 ± 13	234 ± 25	72 ± 3
Dexamethasone + CPT-cAMP	370 ± 21[a]	479 ± 35[a]	77 ± 2

Note: Primary hepatocytes were cultured for 24 h and then dexamethasone (1 μ*M*) and CPT-cAMP (50 μ*M*) were added and incubation continued for an additional 24 h-period. At the end of this period, BCKDH was extracted from hepatocytes and its activity determined. Basal activity was determined before incubating the enzyme with a broad-specificity phosphoprotein phosphatase and the total activity was determined after incubating the enzyme with phosphatase (25 munits/mg protein for 60 min at 37°C). The results are mean ± S.E. of eight hepatocyte cultures. Activity state = (basal activity × 100)/total activity.

[a] $P < 0.01$, control vs. dexamethasone + CPT-cAMP.

From Chicco, A. G., Adibi, S. A., Liu, W. Q., Morris, S. M., and Paul, H. S., *J. Biol. Chem.*, 209, 19427, 1994. With permission.

plus CPT-cAMP (Table 1), indicating that these agents did not increase the enzyme activity by altering the phosphorylation state of the enzyme complex. These results show that dexamethasone and CPT-cAMP-induced increases in the rate of flux indeed reflected increased BCKDH activity.

Changes in BCKDH Subunit Mass

In the above study, an increase in the total activity of BCKDH in hepatocytes cultured in the presence of dexamethasone plus CPT-cAMP strongly suggested an increase in the amount of BCKDH. This possibilty was investigated by quantitative Western blot analysis. The amounts of E1α, E1β, and E2 subunits of BCKDH detected by immunoblot analysis were significantly higher in hepatocytes cultured with dexamethasone and CPT-cAMP (Figure 3, panel A). The greatest increase was observed in the amount of E1α subunit, followed by the E1β subunit, and least for the E2 subunit of BCKDH (Figure 3, panel A). Quantitation by densitometric analysis revealed that there were 3.25-, 2.90-, and 1.60-fold increases in the abundance of E1α, E1β, and E2 subunits of BCKDH, respectively in hepatocytes cultured in the presence of dexamethasone plus CPT-cAMP (Figure 3, panel B). It is clear from these results that the expression of three subunits of BCKDH in response to dexamethasone plus CPT-cAMP, although coordinated, is not stoichiometric. Consequently, the maximum increase in BCKDH complex activity in response to dexamethasone plus CPT-cAMP could be expected to correspond to the subunit of BCKDH with the least increase in protein mass. Among the three subunits of BCKDH, the least increase in protein mass was observed for the E2 subunit. As increases in both the flux through BCKDH and the total activity of

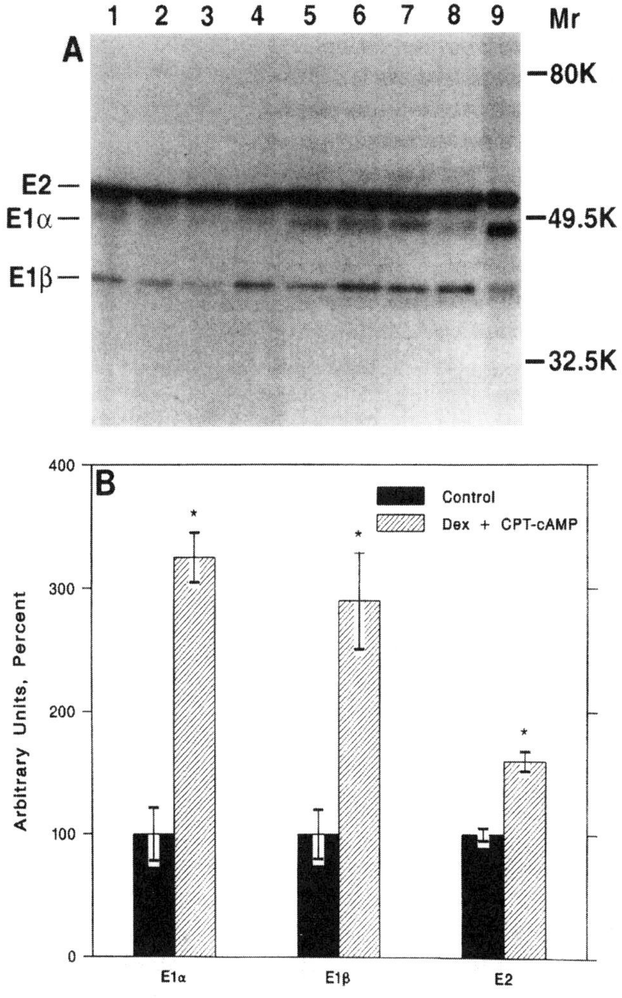

FIGURE 3. Western-blot analysis of $E_1\alpha$, $E_1\beta$, and E_2 subunits of BCKDH isolated from hepatocytes cultured in the absence or presence of dexamethasone plus CPT-cAMP. BCKDH complex was extracted from hepatocytes cultured with or without dexamethasone plus CPT-cAMP and subjected to SDS-PAGE. Resolved proteins were transferred to nitrocellulose membranes and subjected to immunoblot analysis using polyclonal antibodies against rat BCKDH. Subunits of BCKDH were detected by enhanced chemiluminescence using peroxidase conjugated goat anti-rabbit antibody. **A.** lanes 1–4 represent subunits of BCKDH from 4 independent hepatocyte cultures grown in the absence of dexamethasone and CPT-cAMP; lanes 5–8 represent subunits of BCKDH from 4 independent hepatocyte cultures grown in the presence of dexamethasone (1 μM) plus CPT-cAMP (50 μM); lane 9 represents purified BCKDH. Molecular weight markers are shown on the right. **B.** quantitative densitometric analysis of Western blot. Values are presented as mean ± S.E. of hepatocyte cultures from four animals and are expressed as percent of control. Induction in the relative abundance of either $E_1\alpha$, $E_1\beta$, or E_2 subunits of BCKDH by dexamethasone plus CPT-cAMP was statistically significant (*; $P < 0{,}01$). (Reproduced from Chicco, A. G., Adibi, S. A., Liu, W. Q., Morris, S. M., and Paul, H. S., *J. Biol. Chem.*, 269, 19427, 1994. With permission.)

BCKDH complex in response to dexamethasone plus CPT-cAMP were generally twofold, these results indicate that levels of the E2 subunit are limiting for the expression of BCKDH activity.

Changes in Abundance of mRNAs Encoding BCKDH Subunits

To determine whether the effect of dexamethasone and CPT-cAMP reflected regulation at a pretranslational step, Northern analysis was performed. This analysis demonstrated that the mRNA levels for E1α, E1β, and E2 subunits of BCKDH were elevated in hepatocytes cultured in the presence of dexamethasone plus CPT-cAMP (Figure 4, panel A). The greatest increase in relative mRNA abundance occurred for the E1β subunit (21.7-fold), followed by the E1a subunit (7.4-fold), and least for the E2 subunit (4.8-fold) of BCKDH in hepatocytes cultured in the presence of dexamethasone plus CPT-cAMP for 24 hours (Figure 4, panel B). This difference in the relative abundance of mRNAs was not due to differences in RNA loading since the abundance of glyceraldehyde-3-phosphate dehydrogenase (GAPDH) mRNA was not affected by dexamethasone and CPT-cAMP (Figure 4, panel A). Like the increases in the subunit masses in response to dexamethasone plus CPT-cAMP, the increase in mRNA levels for BCKDH subunits was coordinated, but nonstoichiometric.

It is notable that the mRNAs encoding the three subunits of BCKDH differed not only in their relative levels in the absence of hormones but also in the extent of their induction by dexamethasone plus CPT-cAMP. In comparison, as discussed above, a different pattern of nonstoichiometric expression of BCKDH subunit genes was noted in rat liver in response to dietary protein. Thus, it appears that a lack of stoichiometric expression is a general feature for the BCKDH subunit genes.

mRNA Levels and BCKDH Activity in Hepatocytes Relative to Rat Liver

In order to shed light on the possible basis for the apparently large differences in induced levels of mRNAs encoding the individual BCKDH subunits, BCKDH mRNA levels in cultured hepatocytes were compared to those in adult rat liver. In untreated control hepatocytes, the basal levels of mRNA for the three subunits of BCKDH were significantly lower than in adult rat liver (Figure 5). Furthermore, reductions in basal levels of mRNA were different for each subunit. The levels of E1α, E1β, and E2 mRNAs were 72, 16, and 32%, respectively, of *in vivo* levels (Figure 5). When hepatocytes were cultured with dexamethasone and CPT-cAMP, levels of mRNAs for E1α, E1β, and E2 subunits of BCKDH increased 5.3-, 3.4-, and 1.6-fold, respectively, above the level found in rat liver (Figure 5). Thus, apparent variations in induced levels of BCKDH mRNAs reflect not only actual differences in final induced mRNA levels among the subunits, but also significant differences in the initial basal mRNA levels.

The total activity of BCKDH in hepatocytes cultured without dexamethasone plus CPT-cAMP was 24% of that in the liver of an intact animal (Table 2).

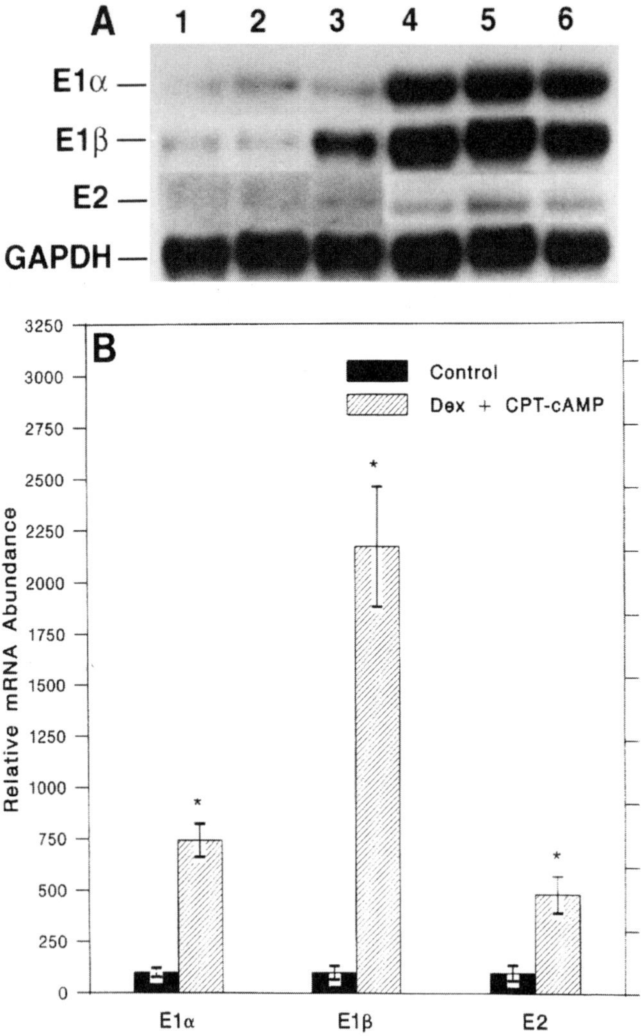

FIGURE 4. Northern blot analysis of mRNA for $E_1\alpha$, $E_1\beta$, and E_2 subunits of BCKDH and for GAPDH from hepatocytes cultured in the absence or presence of dexamethasone plus CPT-cAMP for 24 h. **A.** Total hepatocyte RNA was extracted and subjected to Northern blot analysis using ^{32}P-labeled cDNAs for $E_1\alpha$, $E_1\beta$, and E_2 subunits of rat BCKDH and for GAPDH. Lanes 1–3 represent mRNA from three independent hepatocyte cultures grown in the absence of dexamethasone and CPT-cAMP; lanes 4–6 represent mRNA from three independent hepatocyte cultures grown in the presence of dexamethasone (1 μ*M*) plus CPT-cAMP (50 μ*M*). **B.** Densitometric analysis of Northern blot. mRNA levels of each sample were normalized to the abundance of GAPDH mRNA. Values are mean ± S.E. of hepatocyte cultures from three animals. The induction of mRNAs for either $E_1\alpha$, $E_1\beta$, or E_2 subunits of BCKDH complex by dexamethasone plus CPT-cAMP was statistically significant (*; $P < 0.01$). (Reproduced from Chicco, A. G., Adibi, S. A., Liu, W. Q., Morris, S. M., and Paul, H. S., *J. Biol. Chem.*, 269, 19427, 1994. With permission.)

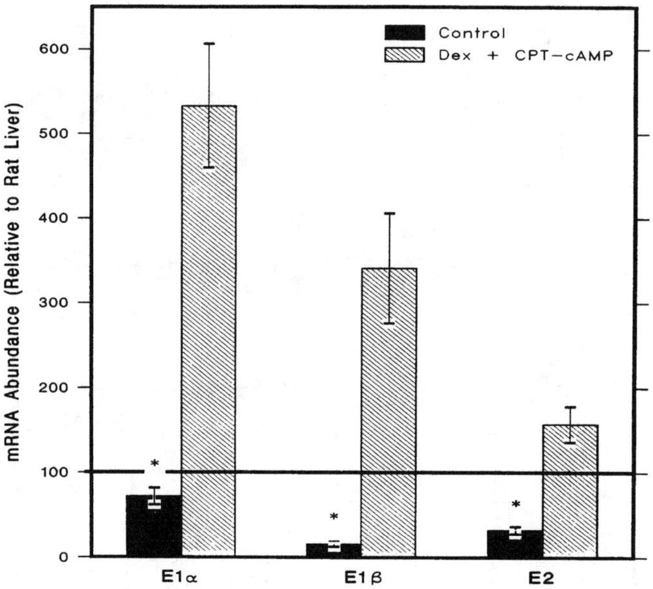

FIGURE 5. BCKDH mRNA abundance in control and hormone-treated hepatocytes relative to that in rat liver. BCKDH mRNA levels in cultured hepatocytes (data from Figure 4) are expressed relative to the average values determined for three adult rat livers (equal to 100; horizontal line). Values are mean ± S.E. of hepatocyte cultures from three animals. The levels of mRNA in control hepatocytes were significantly lower than the mRNA levels in rat liver (*; $P < 0.05$ for $E_1\alpha$ and $P < 0.01$ for $E_1\beta$ and E_2 subunits). The levels of mRNA in hepatocytes cultured with dexamethasone and CPT-cAMP were significantly higher than the mRNA levels in rat liver or in control hepatocytes ($P < 0.05$ for E_2 and $P < 0.01$ for $E_1\alpha$ and $E_1\beta$ subunits). (Reproduced from Chicco, A. G., Adibi, S. A., Liu, W. Q., Morris, S. M., and Paul, H. S., *J. Biol. Chem.*, 269, 19427, 1994. With permission.)

Although the total activity increased by twofold when hepatocytes were cul tured with dexamethasone plus CPT-cAMP, it was only half the activity found in the intact liver (Table 2).

From the present results it appears that glucocoticoids and hormones which act via cAMP play a role in hepatic expression of BCKDH *in vivo*. This is suggested by drops below *in vivo* levels in the abundance of mRNAs encoding BCKDH subunits and in total BCKDH activity in hepatocytes cultured without hormones (Figure 5; Table 2). The fact that dexamethasone and CPT-cAMP induced mRNAs above *in vivo* levels (Figure 5) suggests that the effect of these agents *in vivo* is balanced by counter-regulatory hormones. The failure of dexamethasone and CPT-cAMP to restore BCKDH activity of cultured hepatocytes to the level found in the intact liver suggests: (a) there is also regulation of BCKDH at translational and post-translational steps, and (b) other hormones are necessary for full expression of BCKDH.

TABLE 2
Total Activity of BCKDH in Cultured Hepatocytes vs. Intact Liver

	Treatment	Total BCKDH Activity nmol/min/g wet weight
Hepatocytes	Control	234 ± 25
	Dexamethasone + CPT-cAMP	479 ± 35
Rat liver		986 ± 14[a]

Note: Primary hepatocytes were cultured for 24 h and then dexamethasone (1 μ*M*) and CPT-cAMP (50 μ*M*) were added to one set of hepatocyte culture and incubation of control and treated hepatocytes continued for an additional 24-h period. At the end of this period, BCKDH was extracted from hepatocytes as well as from the rat liver and its total activity was determined after incubating the enzyme with a broad-specificity phosphoprotein phosphatase (25 m units/mg protein for 60 min at 37°C). The results are mean ± S.E. of eight hepatocyte cultures and three livers.

[a] $P < 0.01$, hepatocytes cultured with dexamethasone + CPT-cAMP vs. rat liver.

Role of On-going Protein Synthesis on BCKDH mRNA Induction

The induction of mRNAs encoding the E1α, E1β, and E2 subunits of BCKDH by dexamethasone plus CPT-cAMP required on-going protein synthesis because this induction was largely abolished in hepatocytes cultured in the presence of cycloheximide. Cycloheximide, however, had no significant effect on the corresponding mRNA levels in hepatocytes cultured without dexamethasone and CPT-cAMP (Figure 6).

Although the precise molecular mechanism(s) by which dexamethasone plus CPT-cAMP increased levels of mRNAs encoding the three subunits of BCKDH is not known, DNA sequence analysis has revealed the presence of several potential regulatory elements in the promotor regions of the human E1α, E1β, and E2 genes which resemble glucocorticoid response and cAMP response elements.[1] This is consistent with the notion that induction of the BCKDH subunit mRNAs may reflect increased transcription of these genes. Finally, the inhibition of induction of BCKDH by cycloheximide indicates either that induction is a secondary response which involves hormonal induction of some mediator protein(s), or that induction requires the sustained expression of a preexixting protein(s) which is short-lived.

In summary, primary cultured hepatocytes provide a valuable experimental model for investigating the expression of genes encoding the subunits of BCKDH. Using this model, it has been shown that a hormonally increased flux through BCKDH is due to increased expression of BCKDH subunit genes. Further studies with this model may determine the molecular mechanisms involved in regulating the expression of these genes.

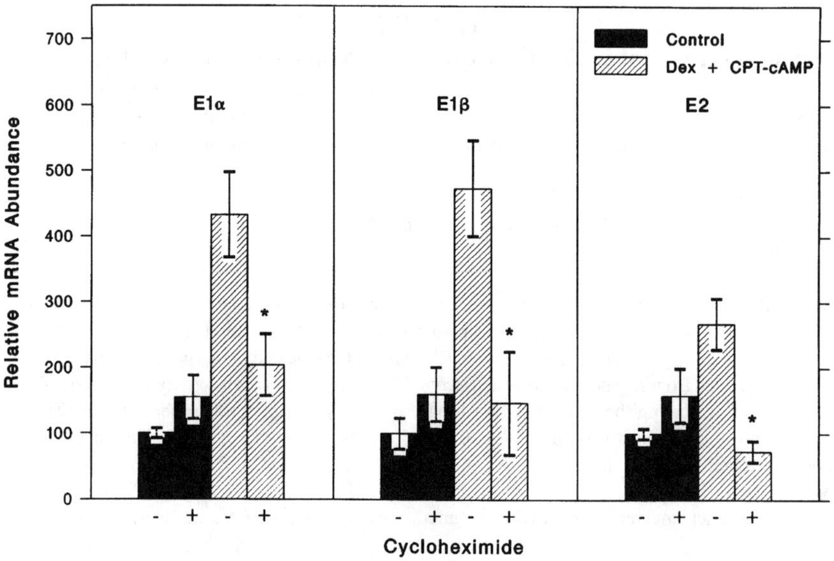

FIGURE 6. Effect of cycloheximide on the induction of mRNA of BCKDH subunits by dexamethasone plus CPT-cAMP. Hepatocytes (0.5×10^5 cells/cm^2) were cultured for 24 h and then dexamethasone (1 μM), CPT-cAMP (50 μM), and cycloheximide (10 μg/ml) were added and incubation continued for an additional 12-h period. Hepatocyte cultures without dexamethasone, CPT-cAMP, and cycloheximide were also maintained for 12 h as controls. After 12 h, total hepatocyte RNA was extracted and subjected to Northern blot analysis using ^{32}P-labeled cDNAs for $E_1\alpha$, $E_1\beta$, and E_2 subunits of rat BCKDH and GAPDH. Blots were subjected to autoradiography and the intensity of each band was determined by densitometry. mRNA levels of each sample were normalized to the abundance of GAPDH mRNA. Values are mean ± S.E. of hepatocyte cultures from four animals. In dexamethasone plus CPT-cAMP-treated hepatocytes, decreases in the relative abundance of mRNAs encoding for the $E_1\alpha$, $E_1\beta$, and E_2 subunits of BCKDH complex in the presence of cycloheximide were statistically significant (*; $P < 0.01$). (Reproduced from Chicco, A. G., Adibi, S. A., Liu, W. Q., Morris, S. M., and Paul, H. S., *J. Biol. Chem.*, 269, 19427, 1994. With permission.)

CONCLUSIONS

Previous studies have concluded that changes in the activity of BCKDH are largely the result of phosphorylation (inactivation) and dephosphorylation (activation) of the enzyme protein. This has resulted in the general view that the genes encoding the BCKDH subunits are unresponsive to metabolic and hormonal perturbations. However, studies reviewed here show that besides covalent modification, regulation of BCKDH is also exerted at the level of gene expression. Since the three catalytic components of the BCKDH complex are present in fixed proportion to each other, coordinated expression of three catalytic components is essential for changes in the activity of the entire complex. Although such a coordinated pattern of expression is apparent during differentiation and development, the expression pattern in intact liver or in primary hepatocyte culture does not appear to be tightly coordinated. The lack of coordinated expression of subunits of BCKDH in relation to the whole complex activity is intriguing. Much is yet to be learned about the regulation of BCKDH gene expression. The tools to explore this exciting and challenging area of research are now available.

REFERENCES

1. **Chuang, D. T. and Shih, V. E.,** Disorders of branched chain amino acid and keto acid metabolism, in *The Metabolic and Molecular Basis of Inherited Disease,* 7th ed., Scriver, C. R., Beaudet, A. L., Sly, W. S., and Valle, D., Eds., McGraw-Hill, New York, 1994, 1239.
2. **Pettit, F. H., Yeaman, S. J., and Reed, L. J.,** Purification and characterization of branched-chain α-keto acid dehydrogenase complex of bovine kidney, *Proc. Natl. Acad. Sci. U.S.A.,* 75, 4881, 1987.
3. **Yeaman, S. J.,** The 2-oxo acid dehydrogenase complexes: recent advances, *Biochem. J.,* 257, 625, 1989.
4. **Cook, K. G., Bradford, A. P., and Yeaman, S. J.,** Resolution and reconstitution of bovine kidney branched-chain 2-oxo acid dehydrogenase complex, *Biochem. J.,* 225, 731, 1985.
5. **Carothers, D. J., Pons, G., and Patel, M. S.,** Dihydrolipoamide dehydrogenase: functional similiarities and divergent evolution of the pyridine nucleotide-disulfide oxidoreductase, *Arch. Biochem. Biophys.,* 268, 409, 1989.
6. **Paxton, R. and Harris, R. A.,** Isolation of rabbit liver branched-chain α-keto acid dehydrogenase and regulation by phosphorylation, *J. Biol. Chem.,* 257, 14433, 1982.
7. **Paul, H. S. and Adibi, S. A.,** Regulation of branched-chain amino acid catabolism, in *Branched-Chain Amino and Keto Acids in Health and Disease,* Adibi, S. A., Fekl, W., Langenbeck, U., and Schauder, P., Eds., S. Karger, Basel, 1984, 182.
8. **Harper, A. E., Miller, R. H., and Block, K. P.,** Branched-chain amino acid metabolism, *Ann. Rev. Nutr.,* 4, 409, 1984.
9. **Wynn, R. M., Davie, J. R., Cox, R. P., and Chuang, D. T.,** Chaperonins groEL and groES promote assembly of heterotetramers ($\alpha_2\beta_2$) of mammalian mitochondrial branched-chain α-keto acid decarboxylase in *Escherichia coli, J. Biol. Chem.,* 267, 12400, 1992.
10. **Adibi, S. A.,** Roles of branched-chain amino acids in metabolic regulation, *J. Lab. Clin. Med.,* 95, 475, 1980.
11. **Goodwin, G. W., Zhang, B., Paxton, R., and Harris, R. A.,** Determination of activity and activity state of branched-chain α-keto acid dehydrogenase in rat tissues, *Methods Enzymol.,* 166, 189, 1988.
12. **Paul, H. S. and Adibi, S. A.,** Mechanism of increased conversion of branched-chain keto acid dehydrogenase from inactive to active form by a medium-chain fatty acid (octanoate) in skeletal muscle, *J. Biol. Chem.* 267, 11208, 1992.
13. **Lawson, R., Cook, K. G., and Yeaman, S. J.,** Rapid purification of bovine kidney branched-chain 2-oxoacid dehydrogenase complex containing endogenous kinase, *FEBS Lett.,* 157, 54, 1983.
14. **Heffelfinger, S. C., Sewell, E. T., and Danner, D. J.,** Antibodies to bovine liver branched-chain 2-oxoacid dehydrogenase crossreact with this enzyme complex from other tissues and species, *Biochem. J.,* 213, 339, 1983.
15. **Williamson, J. R., Walajtys-Rode, E., and Coll, K. E.,** Effects of branched-chain a-ketoacids on the metabolism of isolated rat liver cells. I. Regulation of branched-chain a-ketoacid metabolism, *J. Biol. Chem.,* 254, 11511, 1979.

16. **Corkey, B. E., Martin-Requero, A., Walajtys-Rode, E., Williams, R. J., and Williamson, J. R.,** Regulation of the branched chain α-ketoacid pathway in liver, *J. Biol. Chem.,* 257, 9668, 1982.
17. **Paul, H. S. and Adibi, S. A.,** Effect of carnitine on oxidation of branched-chain amino acids by liver and skeletal muscle, *Am. J. Physiol.,* 234, E494, 1978.
18. **May, M. E., Aftring, R. P., and Buse, M. G.,** Mechanism of the stimulation of branched chain oxoacid oxidation in liver by carnitine, *J. Biol. Chem.,* 255, 8394, 1980.
19. **Paul, H. S. and Adibi, S. A.,** Leucine oxidation and protein turnover in clofibrate-induced muscle protein degradation in rats, *J. Clin. Invest.,* 65, 1285, 1980.
20. **Zhao, Y., Jaskiewicz, J., and Harris, R. A.,** Effects of clofibric acid on the activity and activity state of the hepatic branched-chain 2-oxo acid dehydrogenase complex, *Biochem. J.,* 285, 167, 1992.
21. **Harris, R. A., Paxton, R., Powell, S. M., Goodwin, G. W., Kuntz, M. J., and Han, A. C.,** Regulation of branched-chain α-ketoacid dehydrogenase complex by covalent modification, *Adv. Enz. Regul.,* 25, 219, 1986.
22. **Damuni, Z. M., Merryfield, M. L., Humphreys, J. S., and Reed, L. J.,** Purification and properties of branched-chain α-keto acid dehydrogenase phosphatase from bovine kidney, *Proc. Natl. Acad. Sci. U.S.A.,* 81, 4335, 1984.
23. **Popov, K. M., Zhao, Y., Shimomura, Y., Jaskiewicz, J., Kedishvili, N. Y., Irwin, J., Goodwin, G. W., and Harris, R. A.,** Dietary control and tissue specific expression of branched-chain α-keto acid dehydrogenase kinase, *Arch. Biochem. Biophys.,* 316, 148, 1995.
24. **Solomon, M., Cook, K. G., and Yeaman, S. J.,** Effect of diet and starvation on the activity state of branched-chain 2-oxo-acid dehydrogenase complex in rat liver and heart, *Biochim. Biophys. Acta,* 931, 335, 1987.
25. **Soemitro, S., Block, K. P., Crowell, P. L., and Harper, A. E.,** Activities of branched-chain amino acid-degrading enzymes in liver from rats fed different dietary levels of protein, *J. Nutr.,* 119, 1203, 1989.
26. **Zhao, Y., Popov, K. M., Shimomura, Y., Kedishvili, N. Y., Jaskiewicz, J., Kuntz, M. J., Kain, J., Zhang, B., and Harris, R. A.,** Effect of dietary protein on the liver content and subunit composition of the branched-chain a-ketoacid dehydrogenase complex, *Arch. Biochem. Biophys.,* 308, 446, 1994.
27. **Miller, R. H., Eisenstein, R. S., and Harper, A. E.,** Effects of dietary protein intake on branched-chain keto acid dehydrogenase activity of the rat. Immunochemical analysis of the enzyme complex, *J. Biol. Chem.,* 263, 3454, 1988.
28. **Chuang, D. T., Hu, C. W. C., and Patel, M. S.,** Induction of the branched-chain 2-oxo acid dehydrogenase complex in 3T3-L1 adipocytes during differentiation, *Biochem. J.,* 214, 177, 1983.
29. **Frerman, F. E., Sabran, J. L., Taylor J. L., and Grossberg, S. E.,** Leucine catabolism during the differentiation of 3T3-L1 cells. Expression of a mitochondrial enzyme system, *J. Biol. Chem.,* 258, 7087, 1983.
30. **Hu, C. W. C., Lau, K. S., Griffin, T. A., Chuang, J. L., Fisher, C. W., Cox, R. P., and Chuang, D. T.,** Isolation and sequencing of a cDNA encoding the decarboxylase (E1) α precursor of bovine branched-chain α-keto acid dehydrogenase complex, *J. Biol. Chem.,* 262, 9007, 1988.
31. **Chinsky, J. M., Bohlen, L. M., and Costeas, P. A.,** Noncoordinated responses of branched-chain α-keto acid dehydrogenase subunit genes to dietary protein, *FASEB J.,* 8, 114, 1994.
32. **May, E. E., May, M. E., Aftring, R. P., and Buse, M. G.,** Developmental changes in rat liver branched-chain 2-oxo acid dehydrogenase, *Biochem. J.,* 204, 487, 1982.
33. **Zhao, Y., Denne, S. C., and Harris, R. A.,** Developmental pattern of branched-chain 2-oxo acid dehydrogenase complex in rat liver and heart, *Biochem. J.,* 290, 395, 1993.

34. **Eisenstein, A. B., Strack, I., Gallo-Torres, H., Georgiadis, A., and Miller, O. N.,** Increased glucagon secretion in protein-fed rats: lack of relationship to plasma amino acids, *Am. J. Physiol.,* 236, E20, 1979.
35. **Nebes, V. L. and Morris, S. M.,** Induction of mRNA for phosphoenolpyruvate carboxykinase (GTP) by dexamethasone in cultured rat hepatocytes requires ongoing protein synthesis, *Biochem. J.,* 246, 237, 1987.
36. **Nebes, V. L. and Morris, S. M.,** Regulation of messenger ribonucleic acid levels for five urea cycle enzymes in cultured rat hepatocytes. Requirements for cyclic adenosine monophosphate, glucocorticoids, and ongoing protein synthesis, *Mol. Endocrinol.,* 2, 444, 1988.
37. **Chicco, A. G., Adibi, S. A., Liu, W. Q., Morris, S. M., and Paul, H. S.,** Regulation of gene expression of branched-chain keto acid dehydrogenase complex in primary cultured hepatocytes by dexamethasone and a cAMP analog, *J. Biol. Chem.,* 269, 19427, 1994.
38. **Schmid, E., Schmid, W., Jantzen, M., Mayer, D., Jastorff, B., and Schutz, G.,** Transcription activation of the tyrosine aminotransferase gene by glucocorticoids and cAMP in primary hepatocytes, *Eur. J. Biochem.,* 165, 499, 1987.

Chapter 3

THE CANDIDATE GENE APPROACH IN OBESITY RESEARCH

Carolyn D. Berdanier

CONTENTS

INTRODUCTION

Obesity or excess body fat stores is one of the leading health concerns in wealthy nations of the world. The abundance of food, the availability of machines to perform heavy work, the amount of leisure time available for sedentary activities all contribute to an imbalance in the energy equation. If energy intake exceeds expenditure and there is no way to rid the body of energy in excess of need, this energy is stored as fat. Research has shown that obesity is a risk factor for a variety of degenerative diseases. Research has also shown that obesity is not *simply* due to lifestyle choices. The genetics of the consumer is a key determinant of body fatness.[1] This should come as no surprise to people familiar with farm animals. For decades animal scientists have devoted considerable effort to breeding bovines, pigs, chickens and so forth to maximize rapid weight gain or milk or egg production. Until the advent of biotechnology, these scientists used traditional breeding techniques to achieve success in creating specific characteristics in their animal of choice. If a lean pig with little backfat was desired, selection pressure was placed on this characteristic so that, over time, through selecting breeding pairs with as little backfat as possible, this animal was developed.

Although humans are not manipulated in this way, there are, within certain population groups, subtle social pressures that have similar effects with respect to mate selection.[2] In populations where excess female fatness denotes economic

0-8493-9485-6/96/$0.00+$.50

success of the males, these males tend to select mates having this characteristic. In other populations, the reverse can be observed. In either, clearly, mate selection can have effects on subsequent generations. There appear to be strong genetic factors that determine body fatness. Indeed, given an abundance of food and a sedentary lifestyle, humans will distribute in a typical bell-shaped curve with respect to percent body fat. This distribution will be determined by an interaction of genetic and nongenetic factors.

What is the nature of this interaction? Claude Bouchard in Chapter 5 reports on his many studies on the genetics of human obesity. He and his colleagues have used twins to ask questions about the genetically determined differences in response to exercise, to food intake variation, and to a variety of other treatments. Others[3-5] have also used twins in an effort to elucidate the genetic component in obesity development.

EVIDENCE THAT OBESITY IS AN INHERITED TRAIT

Borjeson[3] studied 40 monozygotic and 61 dizygotic twins and found that early diet had no effect on the body fat content or distribution in these children. With respect to genetic factors, Borjeson reported that the monozygotic twins were more similar in body fatness than the dizygotic twins. Brook et al.[4] measured the skinfold thickness over the triceps and subscapular in 78 monozygotic and 144 dizygotic twins aged 3 to 15 years. These measurements were used to estimate the heritability of such measurements. The overall contribution of the genetic effect was large compared to the contribution of nongenetic effects. Stunkard et al.[5,6] examined single adoptees and twins who were reared by nonrelated adults. In the first study, 540 Danish adoptees, their biological parents and their adoptive parents were evaluated with respect to their body weight. There was a very strong correlation between the adoptees and their biological parents ($P < 0.0001$ for the mothers and $P < 0.0002$ for the fathers) but no relationship between the adoptees and their adoptive parents. Stunkard et al. concluded that genetics are crucial in determining body fatness whereas (under normal circumstances of an abundant food supply) the family environment alone has little influence on this characteristic. In a follow-up study,[6] Stunkard et al. examined 93 pairs of monozygotic twins reared apart, 154 pairs of monozygotic twins reared together, 218 pairs of dizygotic twins reared apart and 208 pairs of dizygotic twins reared together. The intrapair correlation coefficients of the monozygotic twins reared apart were 0.70 for females and 0.66 for males. Those for the twins reared together were only slightly higher (0.74 and 0.66) for female and male respectively, while those of the dizygotic twins were significantly lower (0.15 and 0.25 females and males reared apart and 0.33 and 0.27 for females and males reared together). Like Borjeson, 14 years earlier, Stunkard et al. concluded that diet had little effect on body fatness but that one's genetic background was the most important determinant of this body characteristic. Despite the acknowledgment that obesity or body fatness was an inherited characteristic, scientists interested in the genetic trait were not able to identify specific genes or mutations in genes that determined this trait until quite recently. Candidate genes were proposed but in a number of instances these had to be discarded because the altered gene products appeared to malfunction only as a response to a prescribed set of dietary manipulations or hormonal perturbations. Progress was made nonetheless because scientists had available a number of strains of rats and mice that became spontaneously obese. Among the more important models were the obese Zucker rat and the ob/ob mouse.[7] Both strains inherit their obesity as an autosomal recessive trait. Another mouse, the viable yellow$^{(Avy)}$ mouse inherits its obesity as an autosomal dominant trait. Obesity in the ob/ob mouse has been mapped to chromosome 4 while in the A^{vy} mouse obesity has been mapped to chromosome 2.[7] The mouse named fat inherits its obesity as a mutation on chromosome 8 and the trait in the mouse strain called tubby (tub) has been mapped to chromosome 7. The obese trait in the Zucker rat has been mapped to chromosome 5, but the Zucker diabetic rat and the various strains developed as mutants of the Wistar strain (Koletsky, SHR/cp, LA/N/cp) have not been mapped as yet. A new mouse hybrid AKR/J X SWR/J, has been developed that becomes obese when fed a high fat diet but not when fed a standard stock diet.[8] This responsiveness has been mapped to a locus distal to the db gene on chromosome 4. Osborne-Mendel rats similarly become obese when fed a high fat diet. There may be some interspecies homology in this trait as in the other traits listed above. By comparing physiological characteristics noting similarities and differences, such homologies can be proposed. This then is at the heart of the candidate gene approach in the study of the genetic basis for obesity development.

HYPERLIPOGENESIS

Hyperlipogenesis is common in all of the animal homologs of human obesity.[9,10] Both hepatic and peripheral pathways are more active in the obese phenotype than in the lean phenotype. The rates may differ slightly from one strain to another, but when compared to the lean phenotype within the same strain, the lipogenic rate is substantially increased. Enzymes in the fatty acid synthesis pathway, the pentose shunt, glycolysis, glycogen synthesis, and citric acid cycle are all more active in genetically obese animals than in their lean counterparts. For example, Guichard et al. have reported that the fatty acid synthetase complex is more active in obese Zucker rats than in lean Zucker rats and that this increased activity is due to overexpression of the fatty acid synthase gene at the level of transcription.[11] Fatty acid esterification is likewise increased in the genetically obese[12] as is glucose transporter gene expression[13] and the NADP-linked enzymes.[14] In addition, many of these obese homologs are hyperinsulinemic and evidence peripheral insulin resistance.[9,10] As insulin is important for glucose transport and metabolism, the observations of hyperinsulinemia and hyperlipogenesis are not surprising. Insulin functions in several ways. As a facilitator of glucose uptake by cells, it also serves to stimulate the expression of genes encoding the various enzymes, receptors, and transporters that are needed for glucose disposal. In adipocyte culture (3T3-L1 cells), insulin is an absolute requirement for the expression of virtually every enzyme of the lipogenic pathway.[15,16] Insulin does this by stimulating the dephosphorylation of the CCAAT/enhancer binding protein α, β, and δ. Insulin's action occurs as it is bound to the insulin receptor on the plasma membrane. The steps that occur between this binding and the dephosphorylation of the CCAAT/enhancer binding proteins are unclear at this time. Furthermore, we do not know whether this mechanism of insulin action on gene expression occurs *in vivo* in all cell types. Caloric restriction of preobese rats does prevent the hyperinsulinemia and loss of β cell insulin,[17] but does not alter the gain of fat tissue.[18] Obese Zucker rats restricted to the intake of food consumed by their lean siblings weigh less than their uncontrolled obese cohorts but have the same percent body fat. In other words, although food restriction resulted in a decreased body weight gain, the body composition in terms of fat storage was not affected.

A mutation in a spontaneously obese mouse called "fat" mapped to chromosome 8 results in high levels of circulating proinsulin.[19] The mutation is one which results in an aberrant carboxypeptidase E, the enzyme in the pancreatic β cell which catalyzes the proinsulin processing step. This mouse becomes diabetic as well as obese. The proinsulin is significantly less active at the target tissues than is insulin, yet lipogenesis and fat storage proceeds at an accelerated rate.

HYPERPHAGIA

Another characteristic of genetically obese animals is hyperphagia.[20] The increased food intake by these animals compared to their lean siblings could explain both the hyperinsulinemia and the hyperlipogenesis described above. Typical rodent diets are high in carbohydrate and low in fat. Thus, an animal that is hyperphagic will consume ample amounts of carbohydrate which (a) stimulates insulin release and (b) provides ample substrate for fatty acid synthesis. This hyperphagic characteristic suggests that the genetic fault in these animals may reside in the system which regulates feeding behavior. Either the signals for hunger or those for satiety may be aberrant. In addition, the signals might be correct but the receptors for those signals might be aberrant. The regulation of food intake is discussed in detail by White et al. in Chapter 4.

Several key discoveries led to the hypothesis that in part obesity could be due to one or more genetically aberrant components in the food intake regulation system. First were the discoveries of bloodborne factors that regulated feeding by Hervey[21] and others[22-24] using parabiosed rats or mice (Figure 1). Parabiosis is a technique where two rats or mice are joined together at the skin through the abdominal area. Using genetically lean and obese pairs, it was found that the obese partner decreased his food intake with subsequent reductions in insulin, glucose, and body fat while the lean partner was unchanged.[27] If two normal animals are joined together and one is force-fed to mimic the hyperphagia of the genetically obese animal, the partner of the force-fed animal

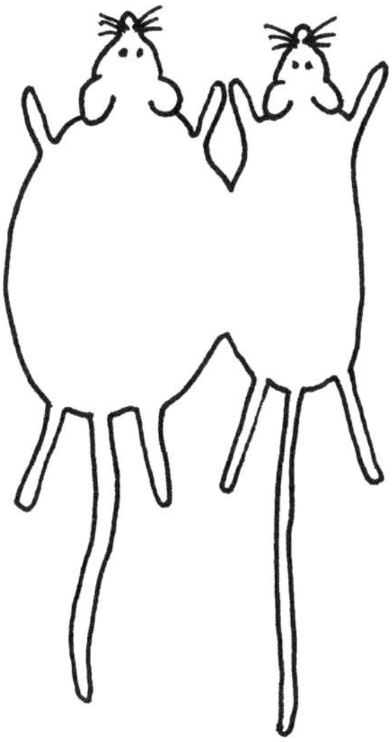

Changes observed:
Lean - no change
Obese - ↓ Food intake
↓ FA synthesis
↓ Insulin
↓ Glucose

FIGURE 1. Schematic representation of a typical parabiotic experiment.

ceases to eat and in fact could starve to death unless the force feeding of the partner was stopped.[23] Both of these experiments suggested that there were blood borne factors that determined feeding behavior. Since much of feeding behavior is determined by the brain, particularly the hypothalamus, there has developed a strong research effort to elucidate genetic aberrations that could explain the hyperphagia of the genetically obese animal. Several of the major investigators in this field explored the notion that genetic obesity is due to a defect in the sympathetic nervous system.[24-27] A number of hormones and their receptors were proposed. These included the glucocorticoids,[27] corticotropin releasing factor,[28] the adenosine receptor,[29,30] adipsin,[31] a variety of neurotransmitters and their receptors,[32-35] and a new signaling hormone called leptin.[36]

Among the neuropeptides that act to regulate food intake is neuropeptide Y (NPY). This is one of the most abundant of the neurotransmitter peptides and, not only does it serve as a satiety signal, but also modulates the activity of a wide variety of cells including those in the endocrine system. Elevated neuropeptide Y levels in the arcuate nucleus of young obese Zucker rats have been reported.[32] The arcuate and paraventricular nuclei form a local network in the hypothalamus and probably constitute an important anatomical component of the feeding/satiety center in this region of the brain. That NPY levels were elevated even before discernible obesity developed in these genetically obese rats, suggested one of two possible scenarios: (1) that elevated NPY levels mean that there is an aberrant receptor for its signal, or (2) that NPY release is aberrant. In either instance NPY levels would increase. In scenario #1, lack of appropriate peptide–receptor binding, the peptide–receptor feedback signal to the NPY synthesis site would be missing whereas in scenario #2 NPY release is uncontrolled. When NPY is injected into the third ventricle of the brains of normal Sprague-Dawley rats, feeding commences.[33] When NPY levels drop, feeding stops. In obese Zucker rats the increase in arcuate nucleus NPY level is related to increased NPY gene expression in this region and hyperphagia is observed.[34] This suggests that NPY levels fail to drop (as in the normal rat) as a response to the infusion of substrates (from food) as they reach the brain. This, in turn, suggests that the genetic aberration is not in the NPY receptor per se but in the signaling system that feeds back and tells the NPY producing cells to stop releasing the peptide. Herzog et al.[35] examined the neuropeptide Y Y1 receptor gene in obese humans and found a variant involving a point mutation in intron 1. However, this variant did not associate with the obese trait. NPY gene expression is controlled at both the transcription and translation levels.[34] Transcription is enhanced by protein kinase C and adenylate cyclase. Drugs such as 12-0-tetradecanoyl phorbol 13-acetate and forskolin, which act via these signaling pathways, can increase the half life of NPY mRNA. Measurements of mRNA stability and assessment of the NPY release mechanism in obese and lean rats have not been reported, but it is possible that genetically determined deviations in these processes may exist and explain the elevated NPY levels in the obese rats. In particular is the possibility that the feedback signal for NPY release is aberrant.

This possibility was explored in the ob/ob mouse by Friedman and colleagues.[36,37] Using positional closing of the mouse ob gene, these researchers reported a 4.5-kb adipose tissue mRNA with a highly conserved 167-amino acid open reading frame. This mRNA is for a secreted protein which these investigators called leptin. A nonsense mutation in codon 105 which expresses 20-fold in the mRNA was found in the obese mouse (ob/ob) and human but not in their lean counterparts. It was suggested that the protein so produced by the adipose tissue served as a signal to the brain with respect to fat depot size and perhaps is the signal for NPY release. An aberrant protein would not be able to serve in this capacity. Impaired expression and functional activity of the β3 and β1 adrenergic receptors in adipose tissue of ob/ob mice has also been reported[38] and these impairments may be related to the impaired signal protein described above. However, an earlier report[39] by this group showed that the ob/ob gene in mice was separate from the fa/fa gene in rats, but that the fa gene in rats was homologous to the db/db gene in mice. The db/db mouse is both obese and diabetic.[7] The ob/ob trait is carried on chromosome 4 and that of the db/db on chromosome 6, while the fa/fa trait in rats maps to chromosome 5.[40] All three are hyperphagic and it is possible that the genetic reasons for this hyperphagia may differ. In the ob/ob mouse the problem may be the signal from the adipose tissue while in the obese Zucker rat and the db/db mouse the problem might rest with the signaling system or the receptors in the brain cells. Research is ongoing to determine where in the signaling system the problem exists in these different animal homologs.

BROWN FAT THERMOGENESIS

In addition to hyperphagia, hyperlipogenesis, and a disturbed insulin–glucose relationship, aberrations in the thermogenic response to cold have been reported[41-43] in genetically obese rodents. Nonshivering thermogenesis occurs in the brown adipose tissue and is stimulated by the catecholamines. Heat is produced by these fat cells when the mitochondria become uncoupled and more energy is released as heat than is trapped in the high energy bond of ATP.

In the newborn, brown adipose tissue has the function of generating heat as part of the body's thermoregulatory process. This is particularly important at this age because newborns lack the insulation provided by the subcutaneous fat store. This fat layer helps conserve body heat. In addition, the newborn's metabolic processes are relatively immature with respect to energy conservation. Heat loss from these processes can be quite large. Thermoregulation, therefore, is quite dependent on the thermogenic capacity of the brown fat. Brown fat thermogenesis is also thought to be a means for ridding the body of excess intake energy. It has been proposed that one of the ways a body maintains a constancy in body weight in the face of inconsistent food energy intake is through this process. On days where intake energy exceeds need, the surplus energy is released as heat via brown fat thermogenesis.

It has been suggested that genetically obese individuals become obese because they are unable to increase their heat production when overfed or when suddenly thrust into a cold environment. It has been hypothesized further, that the genetically obese animal develops the subcutaneous fat pads as insulation against heat loss or gain, thereby circumventing their relative inability to thermoregulate. In order to increase their insulation layer, they must overeat in order to provide the requisite substrates for lipogenesis. Such an hypothesis has some elements of validity. The ob/ob mouse, for example, is unusually sensitive to cold and is incapable of increasing its heat production when suddenly exposed to cold or injected with norepinephrine. These features of the ob/ob mouse precede the development of both hyperphagia and obesity. Since the mobilization of body fuel (induction of lipolysis and glycogenolysis) is not abnormal in the obese ob/ob mouse, it would appear that only the release of heat when these fuels are oxidized is defective.

Impaired thermoregulation has been reported in genetically obese Zucker rats. Lower body temperatures in the neonates and an impaired ability to increase thermogenesis in response to overeating, cold exposure, or to epinephrine, has been reported. It has been suggested that obese humans may also have this defect. When stimulated by cold or infusions of norepinephrine, normal weight subjects increased their heat production while obese subjects did not. Normal weight humans appear to regulate their body fat mass by increasing their brown fat heat production when overfed. Stock and Rothwell[44] developed an animal model to study the effects of overfeeding on brown fat thermogenesis. They showed that when offered a variety of energy-rich snack foods, normal animals increased their heat production to maintain a normal weight. However, this mechanism is far from perfect in that over time these overfed animals did become more fat than did their control-fed littermates. Nonetheless, the observation that brown fat can adjust its heat production in response to fluctuations in the energy intake is a very interesting facet of its function within the body. The mechanism whereby its heat production is increased involves the synthesis and release of an uncoupling protein called UCP or thermogenin. This protein is unique to the mitochondria of brown fat cells. It is not found in any other cell type. This protein is not always present and active. In the starved animal, for example, it cannot be found. This means that heat production by the brown fat or brown fat thermogenesis (BAT) can be switched on and off. Thermogenin synthesis is rapid. Exposure to cold or hyperthyroidism or birth or starvation-refeeding triggers a rapid and marked increase in the messenger RNA for this protein as well as a marked increase in the level of the protein. In most animals, the increase in the level of thermogenin parallels an increase in nonshivering thermogenesis. In lean animals, surplus food intake seems to signal thermogenin synthesis. The regulation of thermogenin synthesis occurs at the level of messenger RNA transcription. Thermogenin is a small protein having 306 amino acid residues. The gene which codes for its synthesis has been isolated and sequenced. The protein completely traverses the inner mitochondrial membrane and has a C-terminal region that projects from the outer surface of the membrane. It works by dissipating the proton gradient that is developed when pairs of reducing equivalents are passed down the respiratory chain to make water. The energy that is developed by the respiratory chain is not captured in the high-energy bond of ATP but is, instead, released as heat. The control of its uncoupling function is through the external C terminal region that contains the nucleotide (ADP) binding site. If the thermogenin binds the nucleotide, it is not available to the F_1F_0 ATP synthase for ATP synthesis. Fatty acids interact with the nucleotide binding site such that ADP binding is increased. These fatty acids are released from the stored triacylglycerols via a hormone stimulated cyclic AMP mechanism. This explains how epinephrine and norepinephrine can stimulate BAT. Epinephrine works by increasing the cAMP levels in the cytosol. This, in turn, stimulates the activity of the cytosolic lipoprotein lipase which serves to release the fatty acids from the stored lipid. Norepinephrine is particularly effective with respect to intracellular fatty acid release in the brown fat. Other lipolytic hormones that work through increasing adenylate cyclase activity and cAMP levels also work to increase BAT. Diets rich in polyunsaturated fatty acids potentiate these hormone effects on BAT. The catecholamines stimulate thermogenesis and do so by activating the cyclic AMP second messenger system. This involves adenylate cyclase and the adenosine receptor. If the adenosine receptor is aberrant[29,30] due to a mutation in the gene which encodes it, then the norepinephrine-induced thermogenesis as well as lipolysis would not occur normally. In turn, if the second messenger system was unresponsive, then the response to cold (which elicits a catecholamine response) or to norepinephrine could be understood. If the adenosine receptor was the site of the mutation one could argue that the hyperphagia might be a compensatory mechanism or a response to a signal not sent from the adipose tissue to the brain. Himms-Hagen in a recent review[41] has attempted to reconcile all of the many hypotheses concerning brown fat thermogenesis, hyperphagia, and the glucostatic theory in the development of obesity.

TABLE 1
Candidate Genes for Obesity

Gene	Why
NPY Signaling System	NPY levels are increased in obese animals, as is hyperphagia.
Adipsin	Defective expression of adipsin found in obese animals.
Glucocorticoid	Adrenalectomy corrects excessive fatness.
Adenosine receptor	Aberrant in obese animals and is related to a poor thermogenic response to cold.
Carboxypeptidase E	One mouse homolog lacks normal enzyme activity and has excessive proinsulin levels.
Leptin	Defective satiety signaling
Thermogenin	Genetically obese rodents do not synthesize this protein very well as evidenced by a reduction in the response to cold.

While insulin is not thought of as a lipolytic hormone, it too has a role in regulating the synthesis and activity of thermogenin and BAT. Both the synthesis of the uncoupling protein and the ability of brown adipose tissue to assist in the regulation of body temperature and dissipate the energy provided by excess food intake is impaired in the insulin-deficient animal. Once insulin is restored, the impairment is corrected. Of interest is the observation that many genetically obese rats and mice are hyperinsulinemic, yet they have impaired BAT. This suggests a genetic error in thermogenin synthesis system.

SUMMARY

A number of candidate genes have been proposed as explanations for obesity development. Some genes have been identified but few have sequenced and mapped. Likely we will find that the obesity phenotype is due to one of several genotypes.

REFERENCES

1. **Bouchard, C.,** *The Genetics of Obesity,* CRC Press, Boca Raton, FL, 1994.
2. **Kissebah, A. H. and Krakower, G. R.,** Regional adiposity and morbidity, *Physiol. Rev.,* 74, 761, 1994.
3. **Borjeson, M.,** The aetiology of obesity in children, *Acta Paediatr. Scand.,* 65, 279, 1976.
4. **Brook, C. G. D., Huntley, R. M. C., and Slack, J.,** Influence of heredity and environment in the determination of skinfold thickness in children, *Brit. Med. J.,* 2, 719, 1975.

5. **Stunkard, A. J., Sorensen, T. I. A., Harris, C., Teasdale, T. W., Chakraborty, R., Schull, W. J., and Schulsinger, F.,** An adoption study of human obesity, *N. Engl. J. Med.,* 314, 193, 1986.
6. **Stunkard, A. J., Harris, J. R., Pedersen, N. L., and McClearn, G. E.,** The body mass index of twins who have been reared apart, *N. Engl. J. Med.,* 322, 1483, 1990.
7. **Coleman, D. L.,** Obese and diabetes: two mutant genes causing diabetes-obesity syndromes in mice, *Diabetologia,* 14, 41, 1978.
8. **West, D. B., Waguespack, J., York, B., and Goudey-Lefevre Price, R. A.,** Genetics of dietary obesity in AKR/JXSWR/J mice: segregation of the trait and identification of a linked locus on chromosome 4, *Mammalian Genome,* 5, 546, 1994.
9. **Bray, G. A. and York, D. A.,** Genetically transmitted obesity in rodents, *Physiol. Rev.,* 51, 598, 1971.
10. **Herberg, L. and Coleman, D. L.,** Laboratory animals exhibiting obesity and diabetes syndromes, *Metabolism,* 26, 59, 1977.
11. **Guichard, C., Dugail, I., LeLiepore, X., and Lavau, M.,** Genetic regulation of fatty acid synthetase expression in adipose tissue: overtranscription of the gene in genetically obese rats, *J. Lipid Res.,* 33, 679, 1992.
12. **Jamdar, S. C. and Cao, W. F.,** Triacylglycerol biosynthetic enzymes in lean and obese Zucker rats, *Biochem. Biophys. Acta,* 1255, 237, 1995.
13. **Penicaud, L., Fere, P. , Assimacopoulos-Jeannet, F., Perdereau, D., Leturque, A., Jeanreneaud, B., Picon, L., and Girard, J.,** Increased gene expression of lipogenic enzymes and glucose transporter in white adipose tissue of suckling and weaned obese Zucker rats, *Biochem. J.,* 279, 303, 1991.
14. **Dugail, I., Quignard-Boulange, A., Liepore, X., Ardouin, B., and Lavau, M.,** Gene expression of lipid storage-related enzymes in adipose tissue of the genetically obese Zucker rat, *Biochem. J.,* 281, 607, 1992.
15. **Kaestner, K. H., Christy, R. J., and Lane, M. D.,** Mouse insulin-responsive glucose transporter gene and trans-activation by the CCAAT/enhancer binding protein, *Proc. Natl. Acad. Sci. U.S.A.,* 87, 251, 1990.
16. **MacDougald, O. A., Cornelius, P., Liu, R., and Lane, M. D.,** Insulin regulates transcription of the CCAAT/enhancer binding protein (C/EBP) a, b and d genes in fully differentiated 3T3-L1 adipocytes, *J. Biol. Chem.,* 270, 647, 1995.
17. **Ohneda, M., Inman, L. R., and Ungen, R. H.,** Caloric restriction in obese pre diabetic rats prevents b cell depletion, loss of b cell GLUT 2 and glucose incompetence, *Diabetologia,* 38, 173, 1995.
18. **Deb, S., Martin, R. J., and Herschberger, T. V.,** Maintenance requirement and energy efficiency of lean and obese Zucker rats, *J. Nutr.,* 106, 191, 1976.
19. **Naggert, J. K., Fricker, L. D., Varlamov, O., Nishina, P. M., Rouille, Y., Steiner, D. F., Carroll, R. J., Paigen, B. J., and Leiter, E. H.,** Hyperproinsulinaemia in obese fat/fat mice associated with carboxypeptidase E mutation which reduces enzyme activity, *Nat. Genet.,* 10, 135, 1995.
20. **Herberg, L.,** Animal models of hyperphagia, *Medicographia,* 8, 4, 1986.
21. **Hervey, G. R.,** The effects of lesions in the hypothalamus in parabiotic rats, *J. Physiol. (Lond.),* 145, 336, 1959.
22. **Harris, R. B. S., Hervey, E., Hervey, G. R., and Tobin, G.,** Body fat content in Zucker rats in parabiosis, *J. Physiol.,* 312, 552P, 1981.
23. **Harris, R. B. S., Martin, R. J., and Bruch, R. C.,** Dissociation between food intake, diet composition and metabolism in parabiotic partners of obese rats, *Am. J. Physiol.,* 268, 874, 1995.
24. **Martin, R. J., Harris, R. B. S., and Jones, D. D.,** Evidence for a central mechanism of obesity in the Zucker fatty rat (fa/fa), *Proc. Soc. Exp. Biol. Med.,* 183, 1, 1986.
25. **Krief, S. and Bazin, R.,** Genetic obesity: is the defect in the sympathetic nervous system? A review through developmental studies in the preobese Zucker rat, *Proc. Soc. Exp. Biol. Med.,* 203, 528, 1991.

26. **Weigle, D. S.,** Appetite and the regulation of body composition, *FASEB J.,* 8, 302, 1994.
27. **Bray, G. A., Fisler, J., and York, D. A.,** Neuroendocrine control of the development of obesity: understanding gained from studies of animal models, *Frontiers Neuroendocrinol.,* 11, 128, 1990.
28. **Smoller, J. W., Truett, G. E., Hirsch, J., and Leibel, R. L.,** The Zucker fatty (fa) gene is not a mutation of coritcotropin-releasing factor, *Am. J. Physiol.,* 264, R12, 1993.
29. **Muzzin, P., Revelli, J-P., Kuhne, F., Gocayne, J. D., McCombie, W. R., Venter, J. C., Giacobino, J-P., and Fraser, C. M.,** An adipose tissue-specific b adrenergic receptor. Molecular cloning and down regulation in obesity, *J. Biol. Chem.,* 266, 24053, 1991.
30. **Vannucci, S. J., Klin, C. M., Martin, L. F., and LaNoue, K. F.,** A_1-adenosine receptor-mediated inhibition of adipocyte adenylate cyclase and lipolysis in Zucker rats, *Am. J. Physiol.,* 257, E871, 1989.
31. **Dugail, I., LeLiepore, X., Quignard-Boulange, A., Pairault, J., and Lavau, M.,** Adipsin mRNA amounts are not decreased in the genetically obese Zucker rat, *Biochem. J.,* 257, 917, 1989.
32. **Beck, B., Burlet, A., Bazin, R., Nicolas, J-P., and Burlet, C.,** Elevated neuropeptide Y in the arcuate nucleus of young obese Zucker rats may contribute to the development of their overeating, *J. Nutr.,* 123, 1168, 1993.
33. **Beck, B., Burlet, A., Bazin, R., Nicolas, J-P., and Burlet, C.,** Unexpected regulation of hypothalamic neuropeptide Y by food deprivation and refeeding in the Zucker rat, *Life Sci.,* 50, 923, 1992.
34. **Lerchen, R. A., Yum, D-Y., Krajcik, R., and Minth-Worby, C. A.,** Transcriptional vs. post transcriptional control of NPY gene expression, *Endocrinology,* 136, 833, 1995.
35. **Herzog, H., Selbie, L. A., Zee, R. Y. L., Morris, B. J., and Shine, J.,** Neuropeptide-Y Y1 receptor gene polymorphism: cross sectional analysis in essential hypertension and obesity, *B. B. R. C.,* 196, 902, 1993.
36. **Halaas, J. L., Gajiwala, K. S., Maffei, M., Cohen, S. L., Chait, B. T., Rabinowitz, D., Lallone, R. L., Burley, S. K., and Friedman, J. M.,** Weight-reducing effects of the plasma protein encoded by the obese gene, *Science,* 269, 543, 1995.
37. **Zhang, Y., Proenca, R., Maffei, M., Barone, M., Leopold, L., and Friedman, J. M.,** Positional cloning of the mouse obese gene and its human homologue, *Nature,* 372, 425, 1994.
38. **Collins, S., Daniel, K. W., Rohifs, E. M., Ramkumar, V., Taylor, I. L., and Gettys, T. W.,** Impaired expression an functional activity of the b3 an b1 adrenergic receptors in adipose tissue of congenitally obese (C57BL/6J ob/ob) mice, *Mol. Endocrinol.,* 8, 518, 1994.
39. **Truett, G. E., Bahary, N., Friedman, J. M., and Leibel, R. L.,** Rat obesity gene fatty (fa) maps to chromosome 5: evidence for homology with the mouse gene diabetes (db), *Proc. Natl. Acad. Sci. U.S.A.,* 88, 7806, 1991.
40. **Truett, G. E., Jacob, H. J., Miller, J., Drouin, G., Bahary, N., Smoller, J. W., Lander, E. S., and Leibel, R. L.,** Genetic map of rat chromosome 5 including the fatty locus, *Mammalian Genome,* 6, 25, 1995.
41. **Himms-Hagen, J.,** Brown adipose thermogenesis and obesity, *Progr. Lipid Res.,* 28, 67, 1989.
42. **Himms-Hagen, J.,** Neural control of brown adipose tissue, *Frontiers Neuroendocrinol.,* 12, 38, 1991.
43. **Milner, R. E. and Trayhurn, P.,** Cold induced changes in uncoupling protein and GDP binding sites in brown fat of ob/ob mice, *Am. J. Physiol.,* 257, R292, 1989.
44. **Stock, M. J. and Rothwell, N. J.,** The role of BAT in adoptive diet-induced thermogenesis, in *Hormones, Thermogenesis and Obesity,* Lardy, H. and Stratman, F., Eds., Elsevier, New York, 1989, 95.

Chapter 4

NUTRITIONAL ASPECTS OF NEUROPEPTIDE GENE EXPRESSION

B. Douglas White, Bin He, Marty H. Porter, and Roy J. Martin

CONTENTS

0-8493-9485-6/96/$0.00+$.50

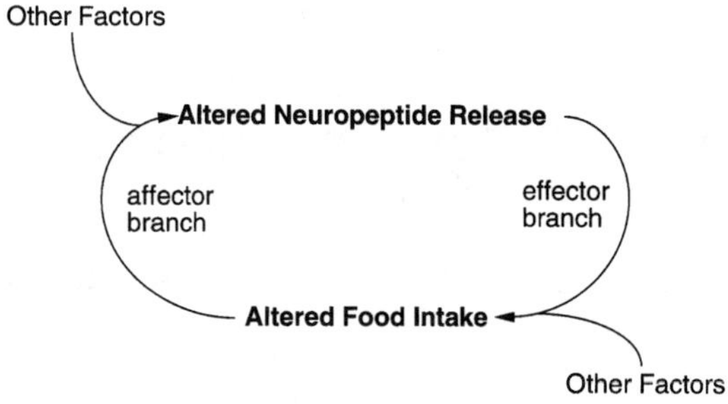

FIGURE 1. A simplified schematic showing a food intake control system involving neuropeptides. The effect of neuropeptides to alter feeding is called the effector branch, while the effect of feeding status to influence neuropeptide release is called the affector branch.

OVERVIEW

A role for nutrients in the regulation of gene expression has been the subject of numerous investigations. Most of these studies have centered on genes that code for proteins important in nutrient processing, such as those involved with digestion, transport, and metabolism. For example, dietary fats have been found to affect the gene expression of gastrointestinal lipase, an enzyme important in the digestion of dietary fat.[1] Amino acid depletion in cultured mammalian cells causes an adaptive increase in amino acid transport. This increase results from the elevated gene expression of a system A transport-associated protein.[2] High levels of dietary carbohydrate increase the gene expression of several lipogenic and glycolytic enzymes.[3,4] Starvation and diabetes increase the gene expression of carnitine palmitoyltransferase, an enzyme important in the oxidation of fatty acids.[5]

From the point of view of a control system, it makes teleological sense that nutrients would regulate pathways important in nutrient processing. In this chapter, we will present evidence that nutrient–gene interactions may also extend to neuropeptides as well. Included among the many proposed functions of neuropeptides is the regulation of food intake and energy balance. Again, from a control system point of view (Figure 1), it makes sense that nutrients would play a role in the gene expression of neuropeptides important in the regulation of food intake.

Though investigations along these lines are just beginning, we will present the current evidence suggesting a relationship between the diet and the regulation of neuropeptides. Specifically, we will examine the role of neuropeptides in the regulation of food intake and energy balance. We will discuss the effects of feeding status on the tissue content and gene expression of these neuropeptides. We will examine the effects of neuropeptides to induce a preference for specific macronutrients. Finally, we will discuss studies that have examined the effects of antimetabolites and dietary macronutrients on the gene expression of specific neuropeptides.

NEUROPEPTIDES

During the past 20 years there has been an explosion of neuropeptide research. To date, more than 40 peptides have been found in the brain.[6] This is attributable, in large part, to advances in biochemical and molecular techniques that have allowed peptides to be isolated and sequenced prior to identification of their physiological function. Because of these technical advances, the current strategy in peptide research has reversed when compared to earlier times.

Early on, researchers began with a physiological function and then searched for a factor that would induce that function. For example, it had been known from the turn of the century that the posterior pituitary contained factor(s) that increased the tone of the uterus, caused milk-secretion and antidiuresis, and had a vasopressor effect. Even after these activities were found to be peptidergic in nature, there was controversy as to the number of peptides responsible for these activities. It is now known the oxytocin is responsible for the tone of the uterus and milk release, while vasopressin is responsible for the vasopressor and antidiuretic effects.[7]

Today, it is more likely that a peptide will be isolated and the molecular biology determined prior to the determination of its physiological function. An example of this is illustrated by the discovery of neuropeptide Y (NPY). NPY was first isolated by chemical means due to its C-terminal amide structure.[8] Molecular methods were then used to translate RNA from human pheochromocytoma cells and identify the mRNA which encoded the protein that was immunoprecipitated with NPY antibodies.[9,10] From the mRNA, a cDNA was synthesized. The cDNA was used to generate a molecular probe that could be used to map the distribution of NPY mRNA-containing cell bodies in the CNS and determine the factors which alter the level of NPY gene expression. Though much is known about the biochemistry and molecular biology of NPY, its proposed physiological functions are still being determined. In Lehmann's 1990 review of NPY, over 60 biological effects were cited as being attributable to NPY.[11] Undoubtedly, the biological effects of NPY are dependent upon the species of animal examined, the site or route of administration and the dose.

The preceding illustrates some of the difficulties in relating neuropeptides to specific physiological functions. One neuropeptide can have more than one physiological function. As an example, a cardiovascular researcher may talk about the presser effect of vasopressin, while a kidney researcher may talk about the body fluid effects of antidiuretic hormone (ADH). Because of its multiple functions, the same peptide is still referred to by either vasopressin or ADH depending on the context of its usage. Another difficulty in relating neuropeptides to specific physiological functions is that a specific physiological function may be affected by multiple interacting neuropeptides. For example, adrenocorticotropin (ACTH) release from the anterior pituitary is stimulated

TABLE 1
Neuropeptides That Affect Food Intake

Increases Food Intake	Decreases Food Intake
Neuropeptide Y	Cholecystokinin
Galanin	Bombesin
Opioids (Dynorphin, B-endorphin)	Neurotensin
Growth Hormone-Releasing Hormone	Insulin
	Glucagon
	Enterostatin
	Corticotropin-Releasing Hormone
	Thyrotropin-Releasing Hormone
	Cyclo(His-Pro)
	Anorectin
	Calcitonin
	Calcitonin Gene-Related Peptide
	Oxytocin
	Vasopressin
	Somatostatin
	Vasoactive Intestinal Peptide
	Insulin-like Growth Factor
	Caerulein

by both corticotropin releasing hormone and vasopressin. Together they have a synergistic effect on ACTH release. Thus, there is not the usual one-to-one relationship between a specific neuropeptide and a specific physiological function.

NEUROPEPTIDES THAT AFFECT FOOD INTAKE AND ENERGY BALANCE

With the above caveats in mind, many neuropeptides have been suggested as having a role in the regulation of food intake and energy balance. It may seem obvious that changes in food intake would have a corresponding effect on energy balance. However, short-term changes in feeding (meal-to-meal) usually do not affect body weight. In addition, it is possible to have a change in energy balance without a change in food intake. For example, obese Zucker rats pair-fed to the level of their lean counterparts still become obese.[12,13] This is believed to be due, at least in part, to a decrease in the activity of brown adipose tissue (BAT) secondary to a decrease in sympathetic activity.[14] Rodents have relatively large amounts of BAT. Increased activity of BAT generates heat, using calories that would otherwise be used for ATP production or energy storage. Many of the neuropeptides that affect food intake also alter sympathetic neural signals to BAT. The neuropeptides that affect food intake fall into two categories: those that increase food intake (orexigenic) and those that decrease food intake (anorexic). A list of neuropeptides that have been suggested to alter food intake is shown in Table 1. Under most circumstances,

TABLE 2
Criteria for Neuropeptide to Qualify as Regulator of Normal Feeding Behavior

1. The neuropeptide increases or decreases food intake in the home environment.
2. The change in food intake occurs during the normal feeding cycle, and the behavioral consequences (i.e., grooming, sleeping, exploring, etc.) resemble normal feeding or food-deprivation.
3. The active locus is the site of endogenous neuropeptide release and the location of neuropeptide receptors.
4. Receptor antagonists have the opposite effect on feeding as the neuropeptide.
5. Under- and overfeeding have an appropriate effects on the endogenous status of the neuropeptide, receptor and gene expression.
6. For orexigenic neuropeptides, the increase in feeding is observed even under conditions where the animals must work to obtain food. For anorexic neuropeptides, animals in an operant environment have a decreased food-reinforced response.
7. Feeding is not due to a direct effect on an alternative behavior.

Adapted from Levine, A. S., and Billington, C. J., in *The Science of Food Regulation: Food Intake, Taste, Nutrient Partitioning, and Energy Expenditure,* Bray, G. A., and Ryan, D. H., Eds., Louisiana State University Press, Baton Rouge, 1992, 210–223. With permission.

there appears to be coordination between the regulation of food intake and the regulation of sympathetic activity to BAT.[15] In general, neuropeptides that increase food intake, decrease sympathetic activation of BAT. Likewise, neuropeptides that decrease food intake, increase sympathetic activation of BAT. Thus, there appears to be a reciprocal relationship between changes in food intake and changes in sympathetic neural activation of BAT.[15] (For a general review of neuropeptides and food intake regulation refer to Morley et al.,[16] Morley et al.,[17] Morley,[18] and Bray.[19])

Though many neuropeptides have been reported to affect feeding, it is desirable to determine whether these factors are natural regulators of hunger and satiety. Levine and Billington have proposed several criteria for a neuropeptide to qualify as a regulator of normal feeding behavior.[20] These criteria, listed in Table 2, can be used as a guide to distinguish specific feeding behavior from nonspecific changes in food intake.

ANOREXIC NEUROPEPTIDES

Some questions should be kept in mind when interpreting results of anorexic agents. These are as follows: what is the behavioral specificity of the effect? Feeding could be decreased because of the specific inhibition of specific elements of feeding response. Does the agent affect other aspects of behavior that would affect feeding? Feeding could be decreased because of a general malaise of the animal. In addition, the arousal level of the animal could be decreased or the mechanics of chewing and swallowing could be inhibited. These responses could lead to a decrease in feeding. Since neuropeptides can have multiple effects, their specificity may be dependent as much on the site of administration and the dose used, as on the peptide itself.[16]

Over a dozen peptides have been shown to inhibit food intake (Table 1). In addition to being produced in the brain, several of the anorexic neuropeptides are also hormones synthesized in and secreted from the gastrointestinal tract or related structures. These are referred to as gut–brain peptides. Below is a brief description of some of the neuropeptides that have been suggested to inhibit food intake. This is not intended to be a complete listing, rather is a survey of some of the neuropeptides that have received attention in the literature.

Gastrointestinal Peptides

Cholecystokinin (CCK)

CCK was first isolated from porcine gastrointestinal tract as a 33 amino acid peptide.[21] CCK has also been shown to be produced in several brain areas as determined by the presence of CCK mRNA.[22] The predominant form of CCK in the brain is the N-terminal octapeptide,[23] which has potent biological activity in feeding.[24] Peripheral as well as intracerebroventricular (ICV) injections of CCK have been shown to inhibit food intake. The peripheral action of CCK appears to be mediated via the vagus nerve with involvement of the area postrema, nucleus tractus solitarius, and paraventricular nucleus (PVN).[25-28] (The reader is referred to Silver and Morley's review of CCK and feeding.[29])

Bombesin

Bombesin is a 14 amino acid peptide first isolated from the skin of the European frog (*Bombina Bombina).*[30] It is found extensively throughout the mammalian gastrointestinal and central nervous system.[31] Cell bodies containing bombesin have been found in the paraventricular nucleus of the hypothalamus and the nucleus of the solitary tract in the brainstem.[32] Both of these areas are associated with feeding. Peripheral injections of bombesin decrease feeding under a variety of conditions.[33-35] As opposed to CCK, the response does not appear to be mediated via the vagus.[36] Direct bilateral administration of bombesin into the lateral hypothalamus has also been shown to decrease feeding.[37] Receptor antagonists given in the fourth ventricle increase short-term food intake, suggesting that endogenous bombesin may have a physiological effect on feeding.[38,39] McCoy and Avery have reviewed bombesin and its effect on feeding.[40]

Neurotensin

Neurotensin is a 13-amino acid peptide first isolated from bovine hypothalami.[41] It is widely distributed throughout the central nervous system with high concentrations found in the hypothalamus and nucleus accumbens.[42,43] Large amounts are also found in the small intestine.[42] ICV injections inhibit both starvation-induced and spontaneous feeding.[44] Tolerance to ICV-administered neurotensin does not appear to develop, at least not within 6 days.[45] Bilateral injections of neurotensin into the paraventricular and ventromedial nuclei of the hypothalamus can also reduce feeding, while injections into the nucleus accumbens have no effect.[45,46]

Pancreatic Peptides

Insulin

Insulin is a 51-amino acid peptide synthesized and secreted from the β cells in the islets of Langerhans in the pancreas. Insulin was the first peptide demonstrated to decrease food intake. In 1974, Hatfield et al. demonstrated that direct injection of insulin into the ventromedial nucleus of the hypothalamus decreased feeding.[47] Insulin receptors have been found in various brain areas.[48] Insulin appears to enter the brain through circumventricular organs and the cerebrospinal fluid.[49,50] Increased concentrations of plasma insulin have been associated with increased insulin concentrations in the cerebrospinal fluid.[51] The concentration of insulin in the cerebrospinal fluid is increased after feeding as well as after ICV infusion of glucose.[52] Injection of insulin antibodies in the ventromedial nucleus stimulates nocturnal feeding, whereas injection in the lateral hypothalamus is without effect.[53] This suggests that endogenous insulin at the ventromedial nucleus acts to inhibit feeding. Chronic ICV infusion of insulin has been shown to decrease feeding and body weight.[54] Consistent with this, chronic infusion of insulin peripherally, at doses which do not cause hypoglycemia, has also been shown to inhibit food intake.[55] It has been proposed that insulin serves as a signal relating information about the status of peripheral energy storage to the central nervous system. Some studies have shown that acute peripheral insulin injections increase food intake.[56] In these cases, the increase in feeding may be due secondarily to hypoglycemia or to a reduction in readily utilizable fuels.

Glucagon

Glucagon is a 29-amino acid peptide synthesized and secreted from α cells in the islets of Langerhans in the pancreas. An early report suggested that glucagon could decrease food intake in humans.[57] This finding was soon confirmed in rats.[58] Glucagon is more potent when administered into the hepatic portal vein, where it more closely mimics the physiological situation.[59,60] There appears to be an association between glucagon-induced satiety and a decrease in hepatic glycogen. These findings prompted the suggestion that glucagon's effect is mediated through alterations in carbohydrate metabolism in the liver. In support of this hypothesis, selective hepatic vagotomy, which severs neural input from the liver to the brain, blocks the satiety effect of glucagon.[61] Intraperitoneal injections of glucagon antibodies increase meal size, suggesting that endogenous glucagon may act to decrease meal size.[62] In contrast to these acute effects, chronic ICV administration of glucagon appears to cause an increase in daily food intake.[63]

Enterostatin

Enterostatin is a 5-amino acid peptide formed by the tryptic digestion of procolipase in the intestine. Procolipase is secreted from the pancreas in response to dietary fat. Peripheral enterostatin has been found to decrease food intake.[64,65] The parent compound, procolipase, decreases food intake when added to food pellets, but colipase does not, suggesting that it is the pentapeptide portion of the procolipase that is responsible for the satiety effect.[66] ICV injections of enterostatin have also been shown to inhibit feeding.[67,68] Enterostatin appears to shorten meal duration without affecting the onset of feeding.[68] These data suggest that enterostatin induces satiety rather than inhibits feeding.

Other Peptides

Corticotropin-Releasing Hormone (CRH)

CRH is a 41-amino acid peptide first characterized in sheep hypothalami.[69] CRH is a component of the hypothalamic–pituitary–adrenal axis. As the name implies, CRH causes the secretion of adrenocorticotropin from corticotrophs in the anterior pituitary. In 1982, it was reported that administration of CRH into the lateral ventricle decreases food intake.[70,71] The effect of CRH on feeding is still present after hypophysectomy.[71] This indicates that the decrease in feeding is not mediated through CRH-induced increases in ACTH concentrations or secondary increases in glucocorticoids. Stress can cause the secretion of CRH. Some stresses have been associated with a decrease in feeding. A CRH antagonist, alpha-helical CRH[9-41] has been shown to attenuate the decrease in feeding associated with restraint stress, suggesting that endogenous CRH may play a role in stress-induced anorexia.[72] The mechanism by which CRH acts to inhibit food intake is not known. It has been reported that CRH can cause a conditioned taste aversion.[73] However, another report has suggested that when injected directly into the paraventricular nucleus, CRH is not aversive.[74] Glowa et al. have extensively reviewed CRH and feeding.[75]

Thyrotropin-Releasing Hormone (TRH) and Cyclo(His-Pro)

TRH is a 3-amino acid peptide first isolated from the hypothalami of sheep.[76] It was the first of the hypothalamic releasing factors to be characterized and synthesized. In addition to the hypothalamus, TRH is found throughout the central nervous system and the gastrointestinal tract.[77] Central, as well as peripheral administration of TRH significantly decreases food intake.[78,79] The peripheral effect of TRH appears to be mediated via the vagus nerve.[80] TRH decreases food intake in hypophysectomized animals, suggesting that the effect is independent of thyrotropin or the secondary release of thyroid hormones.[81] TRH is converted by the enzyme, pyroglutamate aminopeptidease to cyclohistidylprolinediketopiperazine or cyclo(His-Pro).[82] Cyclo(His-Pro) has also been found to cause a decrease of food intake.[83] Since cyclo(His-Pro) has been found to be more potent than TRH, it has been suggested that cyclo(His-Pro) may be the active compound responsible for the inhibition of food intake by TRH.

OREXIGENIC NEUROPEPTIDES

As opposed to the myriad of neuropeptides that have been suggested to inhibit feeding, only four classes of neuropeptides are currently suggested to increase feeding (Table 1). As with the anorectic neuropeptides, there are potential problems in interpreting the results of orexigenic agents. Feeding could result from a specific increase in hunger. Alternatively, the increase in feeding could be due to the increased arousal or indiscriminate chewing by the animal. Again, the specificity of the effect is not only related to the neuropeptide itself, but also to the anatomical site of injection and the dose.

NEUROPEPTIDE Y

NPY is a 36-amino acid peptide first isolated from porcine brain.[8] Exogenous NPY administered into the cerebral ventricles or directly into various areas of the hypothalamus elicits intermittent feeding for up to 4 hr.[84-88] The latency of the feeding effect is approximately 10 min. NPY injected into the paraventricular nucleus causes robust feeding.[89] Since the PVN is a major site of NPY-containing nerve terminals,[90] it was suggested that the PVN may be an important site of action of NPY.[87,89] While the PVN is a site of NPY-induced feeding, mapping studies have since suggested that injections in the perifornical region of the hypothalamus give the most robust feeding response.[91,92] Animals do not appear to develop tolerance to NPY. Chronic administration over a period of several days results in sustained overfeeding and increased body fat accumulation.[93-95] Apparently, NPY is able to override endogenous satiety signals, resulting in body fat accretion. This suggests that NPY could have an effect on long-term feeding and energy balance. The obese Zucker rat, a genetic animal model of obesity, has elevated levels of NPY in the paraventricular nucleus.[96,97] It has been suggested that a sustained increase in NPY may contribute to the etiology of obesity in these animals.[95]

GALANIN

Galanin is a 29-amino acid peptide first isolated in porcine intestine.[98] The central administration of galanin increases feeding in satiated rats.[99] The latency of the feeding effect is approximately 5 min with a duration of 15 to 30 min. The effective site of administration has been suggested to be restricted primarily to the PVN, with some effect noted within the amygdala.[100] A subsequent study suggests that the lateral and the ventromedial hypothalamus are also effective sites of action of galanin on feeding.[101] The feeding effect of galanin probably interacts with the noradrenergic system, because galanin-induced feeding is attenuated by alpha-2 adrenergic receptor antagonists.[102] The galaninergic innervation to the PVN is thought to arise from brainstem nuclei, specifically from the A1 and A6 cell groups.[103] However, high levels of galanin mRNA have also been found by *in situ* hybridization in various areas of the hypothalamus.[104]

GROWTH HORMONE-RELEASING HORMONE (GRH)

GRH was originally isolated and characterized as a 44-amino acid peptide from human pancreatic tumors.[105] The effect of GRH on the release of growth hormone is well known. Though the majority of GRH-containing neurons project to the median eminence, GRH nerve terminals have also been found in intra- and extrahypothalamic sites not associated with the portal blood system.[106] In 1985, it was shown that GRH (administered ICV) could increase the food intake in food-deprived rats.[107] This was later extended to free-feeding animals.[108] Peripheral injections of GRH have no effect, indicating a central site of action.[107] At higher doses, GRH appears to inhibit food intake.[109] This inhibition may be related to the observation that food intake is depressed by low doses of GRH when given during the dark cycle.[110] During the dark cycle, endogenous GRH levels are elevated. It is believed that during the dark cycle, even low doses of exogenous GRH, when added to high endogenous GRH levels, may be sufficient to result in an inhibition of food intake.[110] The medial preoptic area and the suprachiasmatic nuclei appear to be the site of action of GRH's effect on feeding.[111] Because these brain areas are involved in the control of circadian rhythms, it has been suggested that GRH may be one of the neurochemicals that mediates feeding rhythms. In support of this, antibodies to GRH inhibit the increase in feeding associated with the onset of the dark period.[112] GRH antibody treatment, however, does not affect feeding at other times of the day. These data suggest that endogenous GRH plays a role in increased feeding at the onset of the dark period.

OPIOIDS

It has been known for some time that opioids can increase food intake.[113] However, it wasn't until the opioid antagonist, naloxone, was found to decrease food intake that opioids were thought to have a physiological role in food intake regulation.[114] The opioid family of peptides include three classes of precursor proteins: proopiomelanocortin (POMC), proenkephalin, and prodynorphin.[115] From these three precursors, a variety of peptides are derived. Complicating matters, there appear to be at least five major types of opioid receptors: mu, sigma, kappa, delta, and epsilon. Pharmacological evidence has suggested that the kappa and epsilon receptor subtypes play major roles in food intake stimulation. Dynorphin, an endogenous kappa agonist, has been shown to increase food intake.[116,117] Antibodies to alpha-neo-endorphin, another putative endogenous kappa agonist, decrease food intake when injected into the VMN.[118] Central administration of nor-binaltorphamine and naltrexone, selective kappa receptor antagonists, decrease nocturnal and glucoprivic feeding.[119] Central administration of beta-endorphin, an endogenous epsilon receptor agonist, stimulates food intake.[120,121] Antibodies to beta-endorphin, on the other hand, decrease food intake when given centrally.[118] The above data suggest a role of dynorphin and beta-endorphin in the regulation of food intake. Levine et al. have extensively reviewed opioids in feeding.[122]

EFFECTS OF FEEDING STATUS ON NEUROPEPTIDE CONTENT AND GENE EXPRESSION

The above illustrates that neuropeptides can influence food intake and energy balance. As part of a feeding control system (Figure 1), these neuropeptides would be included in the effector branch of the control system (i.e., the part which regulates feeding). As part of the affector branch of the system, feeding status would in turn be expected to regulate the activity of these neuropeptides. This has been found to be the case for a number of the neuropeptides.

FOOD DEPRIVATION

NPY is a potent stimulator of food intake.[84-86] Juxtaposed to NPY's effect on food intake, food intake appears to have a role in the regulation of NPY. Food deprivation over a period of several days causes a dose-dependent elevation in NPY levels in the PVN and arcuate nucleus of the hypothalamus.[123] After 1 day of refeeding, NPY levels in the PVN return to the level of controls. Thus, it appears that the NPY content within the PVN is responsive to the feeding status of the animal. In addition, food deprivation increases NPY gene expression in the arcuate nucleus.[124-127] The arcuate nucleus is a major site of NPY-containing cell bodies whose axons terminate within the PVN.[128] Antisera to NPY cause a dose-dependent attenuation in the feeding response to mild food deprivation.[129] Together, these data suggest that food deprivation increases NPY gene expression in the arcuate nucleus, resulting in a greater release of endogenous NPY. A greater release of NPY would be expected to increase the animal's drive to eat.

Contrary to the effect of food deprivation on NPY gene expression, food deprivation causes a decrease in the gene expression of galanin,[126,130] growth hormone releasing hormone (GRH),[131] and POMC, the precursor of beta-endorphin.[126] The significance of these findings is not yet fully understood. It would tend to suggest, however, that these compounds do not play a large role in increased feeding after food deprivation. Perhaps these data imply that different orexigenic neuropeptides have variable control strengths, depending on the feeding condition.

The gene expression of the anorexic neuropeptide, CRH, is also decreased during food deprivation and food restriction.[126] The decrease in CRH gene expression does not appear to be related to feedback inhibition by glucocorticoids, because circulating corticosterone levels are not elevated. This may serve as a compensatory response to the decrease in feeding, attenuating the inhibition of food intake. It makes sense that anorexic neuropeptides would be decreased at a time when hunger is increased.

Hypothalamic levels of bombesin appear to be responsive to feeding status.[132] Food deprivation for 12 hr causes a decrease in the level of bombesin in the hypothalamus. Feeding returns bombesin levels to that of control. Postprandial levels of bombesin are increased in the hippocampus.

AD LIBITUM FEEDING

Food deprivation may represent an unnatural state to an animal. So, how does neuropeptide content and gene expression relate to the normal feeding state? Rats with ad libitum access to food normally eat the majority of their food during the dark period,[133] with robust feeding beginning around the onset of dark. NPY content in the PVN has been found to increase around this

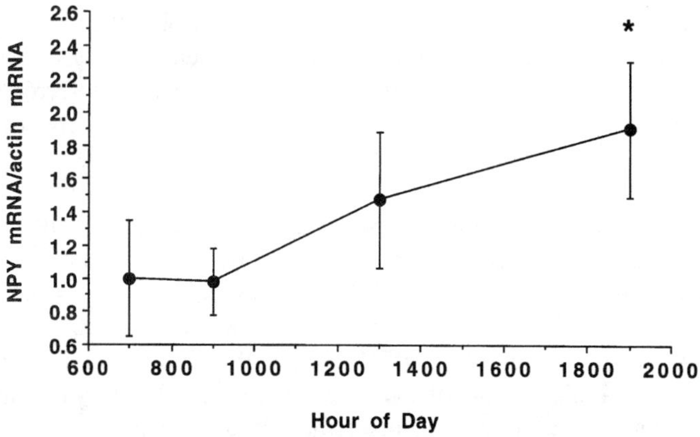

FIGURE 2. Diurnal variation of NPY gene expression in the basomedial hypothalamus of adrenalectomized rats. Rats were adrenalectomized 1 week prior to the experiment. At various times of the day, rats were killed by decapitation and the brains removed. At each time period, rats were killed within 1 hr. The beginning time is indicated on the graph. The basomedial hypothalamus was dissected and the total RNA isolated with RNAzol. Dot blots were made and hybridized with a ^{32}P-labeled riboprobe directed against rat NPY mRNA. Autoradiographs were made and quantified by laser densitometry. Membranes were stripped and rehybridized with a ^{32}P-labeled actin probe. NPY mRNA levels were based on the amount of actin mRNA blotted. Values were normalized to the 7:00 a.m. value. Each point shows the mean ± SEM of 10 animals. * — indicates a significant difference ($p < 0.02$) from the 9:00 a.m. value and tendency for a difference ($p < 0.06$) from the 7:00 a.m. value. (Adapted from White, B. D., Dean, R. G., and Martin, R. J., *Ann. NY Acad. Sci.*, 739, 339, 1994. With permission.)

time.[134,135] There is a decrease in NPY content within the first hour of spontaneous feeding.[134] Moreover, the level of NPY gene expression in the basomedial hypothalamus, site of the arcuate nucleus, increases a few hours prior to the peak of NPY content.[135] We have found NPY gene expression in the basomedial hypothalamus to increase around the onset of the dark period (Figure 2).[136] The level of gene expression was significantly higher at 7:00 p.m., than at 9:00 a.m. Interestingly, this was found in adrenalectomized animals, suggesting that at least a portion of the rhythm is independent of circulating glucocorticoid concentrations. Together, the above data suggest that NPY content and gene expression increase around the time of normal feeding in the rat.

The timing of the NPY increase appears to be related to the time of increased feeding and not the time of day. In a scheduled feeding regimen, food is available to the animal at a set time for only a few hours each day. Under these conditions, it has been found that NPY content and release from the PVN are increased prior to presentation of the food.[137] The levels return to normal within 2 hr of feeding. It is interesting that serum glucocorticoid concentrations also increase prior to feeding under the scheduled feeding regimen.[138] Under ad libitum feeding conditions, serum glucocorticoid concentrations increase around the onset of the dark period, the time when feeding and NPY content and gene expression are increased. Thus, there appears to be a connection between increased serum glucocorticoids, NPY activity, and feeding. The nature of this connection is currently not fully understood. Though some evidence suggests that NPY can stimulate glucocorticoid secretion,[139,140] other evidence suggests a role for glucocorticoids in the increase of NPY content and gene expression.[141-145]

The level of many of the anorectic neuropeptides from the gut is increased after a meal. Within 10 min of feeding, plasma CCK levels increase.[146] In addition, loading of the stomach is associated with an increase in CCK release in the hypothalamus.[147] Thus, it appears that feeding may increase central as well as peripheral CCK levels which in turn act to limit meal size. In an elegant experiment by West and colleagues, it was shown that chronic administration of CCK-8 at the onset of each meal in free-feeding rats causes an average reduction of 44% in meal size.[148] Total food intake is not affected, however, due to a concomitant increase in meal frequency. Additionally, no long-term change in body weight was noted. Thus, CCK may have an effect on short-term feeding (meal-to-meal), but apparently not long-term feeding, which would influence energy balance.

Meal feeding is associated with a rapid increase (within 15 min) in the circulating level of neurotensin.[149,150] Significant circadian variation is found in the hepatic–portal concentration of neurotensin.[151] This variation is linked to the circadian feeding status of the animal. The greatest level of neurotensin is achieved 5 to 10 hr after peak feeding, at a time when animals are in the fasting phase of the circadian feeding cycle.

It is well known that insulin is released in anticipation to feeding. This is the so-called "cephalic phase" of insulin release. Glucose from the meal is a major stimulus of insulin release.[152] Glucagon is released in response to fats and proteins in the diet. Thus, both serum insulin and glucagon concentrations are increased following a well-balanced meal.

As mentioned previously, enterostatin is made from the tryptic digestion of procolipase, and thus, is secreted in response to fat in a meal. Chronic high fat feeding causes progressive increases in the intestinal enterostatin level over time.[153] This increase is paralleled by an elevation in pancreatic lipase activity.

The above examples support the contention that feeding status of the animal can affect the tissue content and gene expression of neuropeptides important in the regulation of feeding. This would be a integral part of the affector branch of a feeding control system (Figure 1).

INFLUENCE OF OREXIGENIC AND ANOREXIC NEUROPEPTIDES ON MACRONUTRIENT PREFERENCES

The above discussion provides evidence that neuropeptides play a role in the regulation of food intake. Furthermore, evidence is presented that supports the hypothesis that food intake affects the regulation of neuropeptides. In addition to the evidence that neuropeptides affect food intake in general, there is abundant evidence to suggest that neuropeptides can affect the intake of specific macronutrients. If an animal is given access to the three macronutrients, fat, protein, and carbohydrate as separate sources, it will eat a certain percentage of each. With a given animal, there is a fairly consistent pattern of macronutrient intake over time. It therefore appears that animals are not eating just for energy but have specific appetites for the macronutrients. Mechanisms appear to exist for the regulation of the intake of individual macronutrients.

In addition to a circadian rhythm of food intake, there is also a rhythm of macronutrient preference. In rats, carbohydrates are preferentially eaten in the early part of the dark period, whereas protein and fats are preferred later in the period.[154] In a 1-hr test, NPY has been found to increase the preference for carbohydrate.[155] As stated above, there is an increase in NPY content and gene expression around the onset of the dark.[134,135] Therefore, it is possible that increased NPY activity around the onset of the dark period contributes to the increase in carbohydrate preference early in the dark period. In support of this hypothesis, there is a positive correlation between the amount of NPY in the parvocellular part of the PVN and the daily amount of carbohydrate consumed.[156] Animals with the highest levels of NPY in the PVN (approximately top quintile) consumed nearly three times more carbohydrate than animals with the lowest levels of NPY in the PVN. Significant positive correlations were also found between carbohydrate consumption and NPY levels in the arcuate nucleus and dorsomedial nucleus. Bilateral injections of an antisense oligonucleotide to NPY mRNA in the arcuate nucleus result in a significant decrease in the carbohydrate intake.[157] Fat intake is also inhibited under these conditions. Others have suggested that under some circumstances NPY may increase fat intake.[158,159] However, when the baseline preference of the animals is accounted for by an analysis of covariance, NPY still appears to cause a preference for carbohydrate.[160]

The timing of the administration of the neuropeptide in relation to the feeding cycle may be important both quantitatively and qualitatively to the induced macronutrient preference. NPY administered early in the dark phase has a stronger effect on carbohydrate intake than it does when administered late in the dark phase.[161] Galanin administered early in the dark phase increases the consumption of both carbohydrate and fat. When administered late in the dark phase, galanin causes a selective increase in the intake of fat, with no effect on carbohydrate intake.[161] Administration of a galanin antagonist to freely feeding rats results in a dose-dependent reduction in fat intake, while protein and carbohydrate intakes are not affected.[162] These data suggest that endogenous galanin may play a role in the regulation of fat intake.

Opioids also appear to have a selective effect on fat intake. After an initial decrease in the intake of all macronutrients, morphine causes a selective increase in fat consumption.[163] Conversely, the opioid antagonist, naloxone, causes a selective decrease in fat selection, suggesting that endogenous opioids act to increase fat intake.[164] The kappa receptor agonists, ketocyclazocine and butorphanol tartrate, cause a selective increase in fat intake, indicating that the effect of opioids to increase fat intake is mediated through the kappa receptor.[165]

Fat in the diet stimulates the release of enterostatin. Enterostatin, in turn, has been found to specifically inhibit the intake of fat.[65] Carbohydrate and protein intakes are not affected by enterostatin. Chronic infusion of enterostatin inhibits body weight gain as well as food intake in Osborne-Mendel rats fed high fat diets,[166] a model of diet-induced obesity. Interestingly, in this study, enterostatin causes a significant increase in the level of circulating corticosterone. Corticosterone has been found to facilitate the inhibition of fat intake by enterostatin.[167] This implies an increased activity of the hypothalamic-pituitary-adrenal axis. There is evidence that CRH can inhibit the intake of fat and protein.[168] The decreased food intake with enterostatin, however, was shown not to be due to a secondary increase in CRH secretion.[166]

Chronic central administration of glucagon has been shown to increase daily food intake.[63] This is in opposition to the anorexic effect of acute administration. The increase in food intake is due to a large increase in the selection of protein, which is enough to compensate for a concomitant reduction in carbohydrate selection.[63] Growth hormone releasing hormone also causes a specific increase in the selection of dietary protein. Daytime injections of GRH into the suprachiasmatic nucleus/medial preoptic area (SCN/MPOA) result in an increase in protein intake without affecting either carbohydrate or fat intake.[169] In addition, injections of GRH antibodies in the SCN/MPOA at the onset of dark cause a selective inhibition of protein intake, suggesting that endogenous GRH contributes to protein intake at night.[170]

CCK and bombesin also affect the intake of specific macronutrients. The short-term response (30 min) to peripheral CCK-8 administration is to decrease fat intake.[171] At higher doses, there is also a slight decrease in protein selection. By 1 hr, no differences in macronutrient selection are discernible. In contrast, injections of an analog of CCK-8, pyroglutamyl-CCK-8, into the ventromedial hypothalamus cause long-term suppression of carbohydrate feeding.[172] The effect of bombesin on macronutrient preference appears to depend largely on the time frame of the observation. Within 30 min, peripheral bombesin administration decreases fat and carbohydrate selection, while between 30 to 60 min protein selection is decreased.[171]

REGULATION OF NEUROPEPTIDES BY MACRONUTRIENTS

The above discussion indicates that neuropeptides affect the intake of specific macronutrients. As part of a feedback control system of food intake, specific macronutrients may play a role in the regulation of neuropeptides. A speculated schematic of a simplified relationship between a neuropeptide and a selected macronutrient is shown in Figure 3. A change in the gene expression of the neuropeptide leads to an alteration in the release of that neuropeptide. The altered release of the neuropeptide causes an alteration in the macronutri-

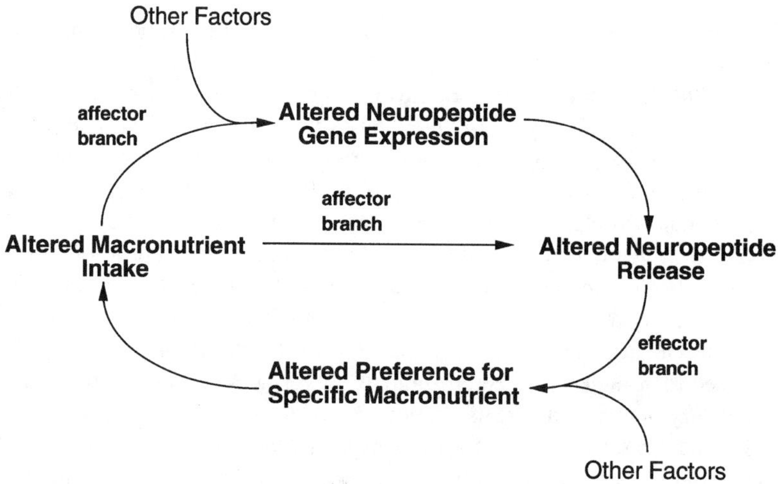

FIGURE 3. Schematic showing the proposed relationship between neuropeptides and macronutrients. It is proposed that an alteration in the gene expression of the neuropeptide leads to a change in neuropeptide release. The altered release of the neuropeptide causes a change in the preference of a specific macronutrient. This leads to an alteration in the animal's intake of that specific macronutrient. The altered macronutrient intake feeds back to change the gene expression of that particular neuropeptide. Release of the neuropeptide may also be affected by the altered macronutrient intake. This feedback changes the turnover of the neuropeptide, readjusting the preference for the specific macronutrient. The result is a feeding control system that tends to stabilize macronutrient intake over time.

ent preference of the animal. Given the availability of the macronutrient, the animal would then alter its consumption of it. The change in macronutrient consumption would then negatively feedback to affect the gene expression and release of the neuropeptide. In addition to neuropeptides, the intake of macronutrients is influenced by numerous other factors, including neurotransmitters, hormones, cytokines, and metabolites. These factors undoubtedly interact with neuropeptides to influence macronutrient preference.

A characteristic of such a system is that it would tend to stabilize macronutrient intake over time. "Under-consumption" of a particular macronutrient would increase the preference for that macronutrient. Conversely, "over-consumption" of the macronutrient would decrease the preference for the macronutrient. This would occur without the need for a classical "set-point." The dynamic equilibrium of the system would set the level of intake of the macronutrient. We have found that the under-consumption of protein leads to an increase in protein intake, whereas over-consumption leads to decreased protein intake (Figure 4). This is consistent with the idea that a feeding control system tends to stabilize macronutrient intake over time.

Few studies have directly examined the effects of macronutrients on the regulation of neuropeptides. As stated above, food deprivation causes a decrease in the gene expression of GRH. The decrease in GRH gene expression

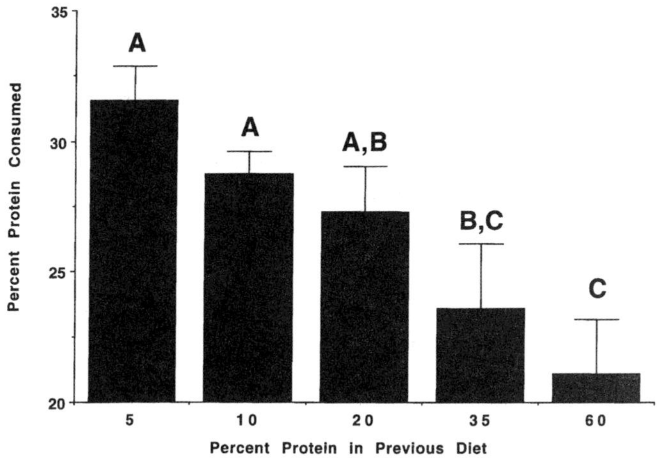

FIGURE 4. Protein selection following diets containing various amounts of protein. Forty-two rats were given ad libitum access to one of five isocaloric diets which varied in the amount protein (5, 10, 20, 35, and 60% casein). The amount of fat was fixed at 10%. The amount of carbohydrate was adjusted to make the diets isocaloric. After the rats were fed these diets for 1 week, they were given a choice between the 5% and 35% casein diets. After 24 hr, the intake of each was determined and the amount of protein consumed was calculated. Each bar shows the mean ± SEM of 8–9 animals. Bars with different superscripts are significantly different ($p < 0.05$).

is normalized after refeeding a balanced diet, a carbohydrate-free diet, or a fat-free diet.[173] However, GRH gene expression remains depressed after refeeding a protein-free diet. If animals are fed isocaloric diets containing various amounts of protein (4 to 12%), GRH gene expression is increased in proportion to the amount of protein. These data indicate that the decrease in GRH gene expression with food deprivation is due to the specific decrease in protein intake.

Most of the studies that have directly examined the effects of macronutrients on the regulation of neuropeptides have centered on NPY. Administration of NPY causes a preference for carbohydrate consumption.[155] Interestingly, blockade of glucose utilization by the antimetabolite, 2-deoxyglucose (2-DG), also increases carbohydrate consumption.[174] It is possible that low glucose utilization stimulates NPY activity. This stimulation in activity would increase the animal's preference for carbohydrate, leading to elevated carbohydrate consumption. Elevated carbohydrate consumption would rapidly increase glucose availability and possibly glucose utilization, thus inhibiting the original signal. 2-DG has been found to increase NPY levels in the arcuate nucleus and to a smaller degree in the suprachiasmatic nucleus.[175] It was concluded that glucose utilization may constitute an important regulatory pathway for NPY production in the hypothalamus. Interestingly, 2-DG treatment did not elevate NPY levels in the PVN, a major site of NPY-containing nerve terminals. Perhaps an increase in the release of NPY offsets a potential increase in NPY content at the site of the nerve terminals.

Neuronal cells, when stimulated, respond by initiating second messenger cascades. The initial signaling steps of this response often activate protein kinases which ultimately result in activation or suppression of gene transcription. In the instance of NPY gene expression, two drugs, 12-0-tetradecanoyl phorbol-13-acetate (TPA) and forskolin, that activate protein kinase C and adenylate cyclase, have been shown to up-regulate NPY gene transcription.[176-183] NPY mRNA increased 21- to 67-fold but translation increased only sevenfold.[182,183] The fold difference in the two processes suggests post-transcriptional regulation, and such regulation has been reported.[184-186] Studies of the transcriptional control of the NPY gene have revealed regions in the proximal promoter that respond to stimulation and this response is that of increasing NPY gene transcription. Several cis-acting elements in the 5′ flanking sequences between –87 and –36 have been identified. In addition, several trans-acting factors can interact with this region.[184,185] Certain unidentified proteins (proteins A and B) interact with the promoter sequences centered around –73; an activator protein 2 (AP-2) has been found that binds sequences centered at –61 and an Sp1-like binding protein reacts with the CT-box sequence located at –51. Lerchen et al. showed that TPA and forskolin do not measurably increase NPY gene transcription in cultured neuroblastoma cells. However, these compounds have effects on a protein (Sp1-like) that seems to mediate the gene response. The CT-box entered at –51 binds this Sp1-like protein.[186] The GC-rich sequence immediately preceding the CT-box is a relatively conserved AP-2 binding consensus sequence. AP-2 is a trans-acting factor that was initially defined by its ability to bind the degenerate site; CCC(/C)N(G/C)(G/C)(G/C) where N represents any nucleotide. It is a helix-span-helix protein that is found in abundance in cultured cells derived from the neural crest. AP-2, however, does not bind to the NPY promoter region. The drugs used by Lerchen et al.[186] seem to stabilize NPY mRNA, effectively lengthening its half life. Whether other compounds, i.e., nutrients, can have a similar effect is the subject of much interest.

The effects of 2-DG on NPY gene expression have been examined with the use of dot/slot blot analyses. Using this technique, no effect of 2-DG on NPY gene expression has been found (Figure 5a).[187,188] However, using *in situ* hybridization, we have recently shown that 2-DG increases NPY gene expression within subregions of the arcuate nucleus (Figure 5b). In the more caudal regions of the arcuate nucleus, 2-DG increases NPY gene expression. NPY gene expression is not affected, however, in the more rostral regions of the nucleus. This finding is similar to the effect of lactation on regional NPY gene expression within the arcuate nucleus.[189] Regional variations in gene expression may explain why differences in NPY gene expression could not be detected by the use of dot/slot blot analysis. With this technique, RNA is isolated from a whole block of tissue, thus pooling subregions together. The nonresponsive rostral region of the nucleus could mask differences in the responsive caudal region. Together, the above data are consistent with a role for glucose utilization in the regulation of NPY activity and gene expression.

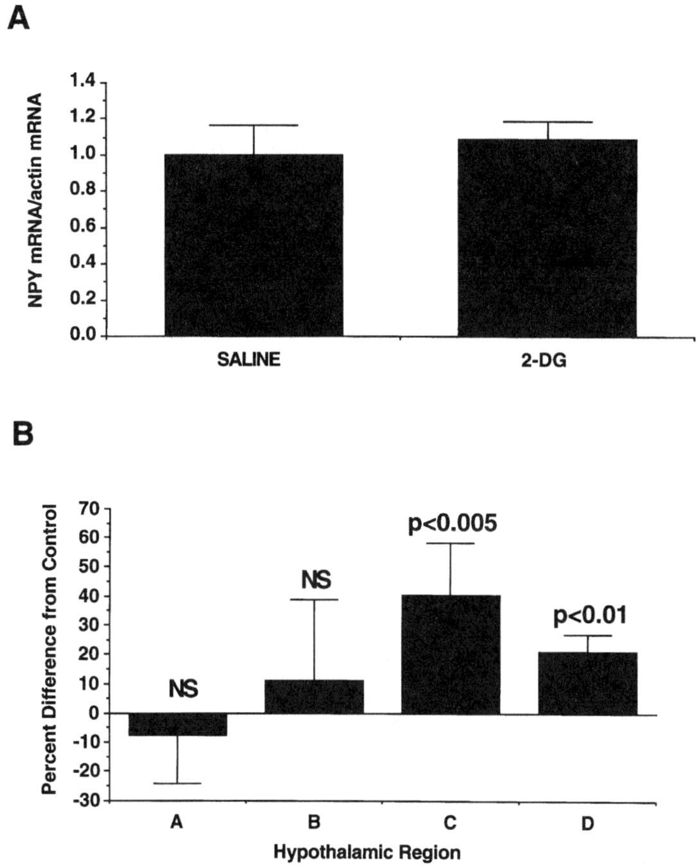

FIGURE 5. (A) NPY gene expression in the basomedial hypothalamus following 2-deoxyglucose (2-DG) treatment as determined by dot blot analysis. Twenty rats were divided into two groups. One group received an IP injection of 2-DG (500 mg/kg), while the other group received an injection of saline. After 2 hr, the animals were killed and the brains removed. (Food was removed during the 2 hr.) The basomedial hypothalamus was dissected and the total RNA isolated with RNAzol. Dot blots were made and hybridized with a ^{32}P-labeled riboprobe directed against rat NPY mRNA. Autoradiographs were made and quantified by laser densitometry. Membranes were stripped and rehybridized with a ^{32}P-labeled actin probe. NPY mRNA levels were based on the amount of actin mRNA blotted. Values were normalized to the saline groups. (B) Regional differences in NPY gene expression of the basomedial hypothalamus of rats following 2-DG as determined by *in situ* hybridization. Pairs of rats were treated with 2-DG as described above. Brains were perfusion-fixed with 4% paraformaldehyde. 30 mm coronal sections were made through the hypothalamus. Sections were hybridized with a ^{32}P-labeled riboprobe directed against rat NPY mRNA. After washing, sections were mounted on slides and an autoradiograph was made. Binding was quantified by image analysis. Four separate rat pairs were independently examined. Sections were grouped based on their rostral-caudal location. With reference to bregma, (A) –1.8 to –2.3 mm, (B) –2.3 to –2.8 mm, (C) –2.8 to –3.3 mm, (D) –3.3 to –4.16 mm. The treatment effect was determined by an analysis of variance, while blocking on autoradiographs. Bars show the average percent difference from control for the four rat pairs.

The effects of dietary macronutrients on NPYergic activity have also been examined. After 14 days, animals fed a high fat/low carbohydrate diet have a greater content of NPY in the parvocellular part of the PVN than do animals fed a high carbohydrate/low fat diet.[190] Neither group, however, was different from animals fed a balanced diet. No difference in NPY content was found in the arcuate nucleus. Accompanying the increased NPY content in the PVN, animals fed the high fat/low carbohydrate diet consumed more calories and gained more weight than did animals on the high carbohydrate/low fat diet. These data may suggest that some factor associated with the high fat/low carbohydrate diet increases NPYergic activity as compared to the high carbohydrate/low fat diet. The pattern of hypothalamic NPY content changes when these diets are fed for longer periods of time (14 weeks). Under these conditions, NPY content is reduced in the arcuate nucleus by both the high carbohydrate and high fat diets as compared to the control diet.[191] NPY content is also decreased in the parvocellular part of the PVN in rats fed the high fat diet compared to the content in rats fed the control diet. NPY content is not affected in other hypothalamic or extrahypothalamic areas. Though animals on the high fat diet are not hyperphagic, they become obese. Rats fed high fat diets are energetically more efficient. That is, the fat in the diet can be transferred to the storage depot with little loss of energy. This transference is a factor separate from the effects of NPY on intake. Animals on the high carbohydrate diet, on the other hand, are hyperphagic, though they have smaller body weights. It was suggested by the authors that low NPY content indicates a decrease in NPY activity and that this occurs as a counterregulatory mechanism in response to hyperphagia (high carbohydrate diet) and increased body fat (high fat diet).

Alone, neuropeptide content data may be difficult to interpret. For example, suppose NPY content is increased in the PVN. This could be due to an increase in the rate of NPY synthesis, suggesting an increase in NPYergic activity. However, it is possible that NPY content is increased because of a decrease in the release of NPY. In this case, NPYergic activity would be decreased. Neuropeptide content itself is not a physiological variable, rather it represents the balance of two opposing physiological variables, i.e., the addition of new neuropeptide into the tissue (synthesis) and the removal of neuropeptide from the tissue (release). The additional determination of gene expression aids in the interpretation of the results by supplying an index of synthesis. This is illustrated in the following.

Rats fed a highly palatable diet become hyperphagic and obese.[192] After 7 weeks on a highly palatable diet (33% condensed milk and 7% sucrose), the content of NPY in the PVN and arcuate nucleus is elevated.[193] Because no difference was found in the hypothalamic NPY mRNA level, the authors concluded that NPY release was decreased in animals fed the palatable diet, leading to the increase in NPY content. Since there was little difference between the carbohydrate and fat content of the control and experimental diets, it was suggested that the difference in NPY content was due to a difference in the palatability of the diets.

Care should also be taken with the interpretation of mRNA data. The gene expression within various nuclei of the hypothalamus may not respond in a manner similar to a particular perturbation. Adrenalectomy has been shown to decrease NPY mRNA level in the basomedial hypothalamus,[143] the site of the arcuate nucleus, whereas adrenalectomy has been shown to increase NPY gene expression in the PVN.[194] Thus, potential differences in gene expression could be overlooked if specific areas are not examined individually.

The gene expression of NPY in the arcuate nucleus has been shown to be altered by the diet. After 1 week, the NPY gene expression in the arcuate nucleus is decreased in rats fed a high fat/low carbohydrate diet as compared to rats fed a high carbohydrate/low fat diet.[195] Rats fed diets of intermediate macronutrient composition produce intermediate levels of gene expression. Despite the change in gene expression, NPY content in the PVN is not affected by the diet. This lack of an effect on the concentration of the peptide could indicate that the synthesis and release of NPY have both decreased to a similar degree. These data suggest that a high fat/low carbohydrate diet decreases NPYergic activity as compared to a high carbohydrate/low fat diet. Interestingly, this conclusion is exactly opposite of the earlier work of Beck and colleagues.[190] The reason for this discrepancy is difficult to determine; however, it may be related to differences in diet composition.

We have systematically examined the effects of macronutrient restriction on the expression of the NPY gene in the arcuate nucleus.[196] In this study, several different isocaloric diets were used. Each macronutrient (carbohydrate, fat, and protein) was individually reduced compared to a standard control diet. After 12 days, NPY gene expression in the arcuate nucleus increased in rats fed low protein diets (Figure 6). No difference in gene expression was found in rats fed low carbohydrate or low fat diets. Additionally, no differences were found in the content of NPY in the PVN. In addition to an increase in NPY gene expression, rats fed the low protein diets were hyperphagic and had an increase in fat accretion,[196] both characteristics of chronic NPY release. These data are consistent with the notion that low protein diets increase the gene expression, synthesis, and release of NPY.

The results of the studies that have examined the effects of macronutrients on NPY activity appear to be somewhat conflicting. One study suggests NPY activity is increased by a high fat/low carbohydrate diet,[190] while another suggests that NPY activity is decreased by a high fat/low carbohydrate diet.[195] Another study suggests NPY activity is decreased by a palatable diet.[193] Finally, we have reported that NPY activity is not affected by dietary carbohydrate or fat, but rather increased by low protein diets.[196] The results of these studies are highly dependent on the exact composition of the diet. Because no two studies used the same diet, a direct comparison of their results is difficult. However, one consistent dietary factor appears to be the level of protein. Within each of these studies, it is the diet with the lowest amount of protein (by weight) that shows the greatest NPY content in the PVN[190,193] or the greatest NPY gene expression in the arcuate nucleus.[195,196]

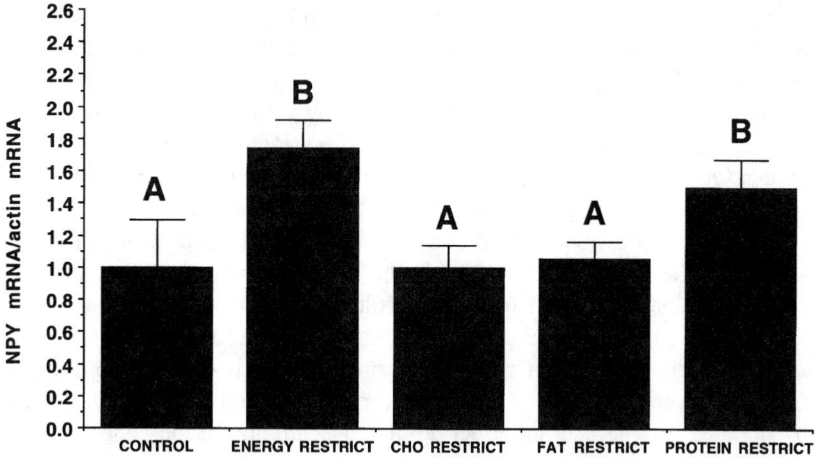

FIGURE 6. Effects of macronutrient restriction on NPY gene expression in the basomedial hypothalamus. Fifty-eight rats were fed a modified AIN-76 diet for 1 week (55% carbohydrate, 10% fat, and 20% protein, by weight). Rats were divided into eight groups. All test diets were isocaloric. One group continued with the control diet. One group continued with the control diet, but their intake was restricted to 50% of control animals. Two groups were fed diets which had only 50% of the carbohydrate as the control diet. The caloric difference was made up with fat in group and protein in the other. Two groups were fed diets which had only 50% of the fat as the control diet. The caloric difference was made up with carbohydrate in group and protein in the other. Two groups were fed diets which had only 50% of the protein as the control diet. The caloric difference was made up with carbohydrate in group and fat in the other. After 1 week, the rats were killed and the brains removed. The basomedial hypothalamus was isolated and the NPY gene expression determined by dot blot analysis as described in the legend of Figure 1. No differences were found diets restricted in alike macronutrients, so the data were combined. Each bar shows the mean ± SEM of 8 to 14 animals. Bars with different superscripts are significantly different ($p < 0.05$). (Adapted from White, B. D., He, B., Dean, R. G., and Martin, R. J., *J. Nutr.*, 124, 1152, 1994. With permission.)

CONCLUSIONS AND PERSPECTIVES

We have presented data showing that several neuropeptides play a role in the regulation of food intake. Though most are produced in the brain, some are also released by the gastrointestinal tract or related areas. These are termed gut–brain peptides. Some neuropeptides are anorexic, decreasing food intake, while other are orexigenic, increasing food intake. Some of the neuropeptides act on a short-term basis, regulating meal-to-meal food intake, while others appear to have long-term effects, influencing body weight. Food intake status appears to play a role in the regulation of many of these neuropeptides. This would be an integral part of the affector branch of a feeding control system. Thus, neuropeptides appear to regulate feeding, while food intake status in turn feeds back to control the regulation of neuropeptides.

More recently, the effects of these neuropeptides on macronutrient selection have been examined. Many neuropeptides have been shown to cause an alteration in the selection of specific macronutrients. This suggests that specific macronutrients may, in turn, play a role in the regulation of neuropeptides. Though such studies are only just beginning, the available evidence supports this idea. More work is needed to further elucidate the effects of macronutrients on the regulation of neuropeptides.

The latest National Health and Nutrition Examination Survey (NHANES) has determined that the prevalence of obesity is increasing in the U.S.[197] One in three people are now overweight. This contributes to cardiovascular disease, diabetes, stroke, and some forms of cancer. Attempts at caloric reduction in order to lose body weight are largely unsuccessful in the long term. This is due in part to a lack of understanding of the basic mechanisms that regulate food intake. A better understanding of how dietary macronutrients regulate neuropeptides that, in turn, regulate feeding would contribute to our overall understanding of these basic mechanisms.

REFERENCES

1. **Wicker-Planquart, C. and Puigserver, A.,** Regulation of gastrointestinal lipase gene expression by dietary lipids, in *Nutrition and Gene Expression,* Berdanier, C. D. and Hargrove, J. L., Eds., CRC Press, Boca Raton, FL, 1993, chap. 3.
2. **Kilberg, M. S., Hutson, R. G., and Laine, R. O.,** Amino acid-regulated gene expression in eukaryotic cells, *FASEB J.,* 8, 13, 1994.
3. **Allred, J. B. and Bowers, D. F.,** Regulation of acetyl CoA carboxylase and gene expression, in *Nutrition and Gene Expression,* Berdanier, C. D. and Hargrove, J. L., Eds., CRC Press, Boca Raton, FL, 1993, chap. 12.
4. **Goodridge, A. G.,** Dietary regulation of gene expression: enzymes involved in carbohydrate and lipid metabolism, *Ann. Rev. Nutr.,* 7, 157, 1987.
5. **Brady, P. S., Barke, R. A., and Brady, L. J.,** Regulation of the 68 kDa hepatic carnitine palmitoyltransferase, in *Nutrition and Gene Expression,* Berdanier, C. D. and Hargrove, J. L., Eds., CRC Press, Boca Raton, FL, 1993, chap. 14.
6. **Evinger, M. J. Q., Chen, C.-L. C., Schachter, B. S., and Roberts, J. L.,** Peptide hormone gene expression in the brain, in *Gene Expression in Brain,* Zomzely-Neurath, C. and Walker, W. A., Eds., John Wiley & Sons, New York, 1985, chap. 2.
7. **Acher, R.,** Biosynthesis, processing, and evolution of neurohypophysical hormone precursors, in *Neurosecretion and the Biology of Neuropeptides,* Kobayashi, H., Barn, H. A., and Urano, A., Eds., Japan Scientific Societies Press, Tokyo, 1985, 11.
8. **Tatemoto, K., Calquist, M., and Mutt, V.,** Neuropeptide Y — a novel brain peptide with structural similarities to peptide YY and pancreatic polypeptide, *Nature,* 296, 659, 1982.
9. **Minth, C. D., Bloom, S. R., Polak, J. M., and Dixon, J. E.,** Cloning, characterization and DNA sequence of a human cDNA encoding neuropeptide tyrosine, *Proc. Natl. Acad. Sci. U.S.A.,* 81, 4577, 1984.
10. **Minth, C. D., Andrews, P. C., and Dixon, J. E.,** Characterization, sequence, and expression of the cloned human neuropeptide Y gene, *J. Biol. Chem.,* 261, 11974, 1986.

11. **Lehmann, J.,** Neuropeptide Y: an overview, *Drug Dev. Res.,* 19, 329, 1990.
12. **Zucker, L. M.,** Some effects of caloric restriction and deprivation on the obese hyperlipemic rat, *J. Nutr.,* 91, 247, 1967.
13. **Martin, R. J. and Gahagan, J.,** Serum hormone levels and tissue metabolism in pair-fed lean and obese Zucker rats, *Horm. Metab. Res.,* 9, 167, 1977.
14. **Moore, B. J., Horwitz, B. A., and Stern, J. S.,** Brown fat thermogenesis and its role in the development of obesity, *Brain Res. Bull.,* 14, 577, 1985.
15. **Bray, G. A.,** Reciprocal relation between the sympatheic nervous system and food intake, *Brain Res. Bull.,* 27, 517, 1991.
16. **Morley, J. E., Levine, A. S., Gosnell, B. A., and Krahn, D. D.,** Peptides as central regulators of feeding, *Brain Res. Bull.,* 14, 511, 1985.
17. **Morley, J. E., Bartness, T. J., Gosnell, B. A., and Levine, A. S.,** Peptidergic regulation of feeding, *Int. Rev. Neurobiol.,* 27, 207, 1985.
18. **Morley, J. E.,** Neuropeptide regulation of appetite and weight, *Endocrin. Rev.,* 8, 256, 1987.
19. **Bray, G. A.,** Peptides affect the intake of specific nutrients and the sympathetic nervous system, *Am. J. Clin. Nutr.,* 55, 265S, 1992.
20. **Levine, A. S. and Billington, C. J.,** Selected criteria for peptides as regulators of feeding: an overview, in *The Science of Food Regulation: Food Intake, Taste, Nutrient Partitioning, and Energy Expenditure,* Bray, G. A. and Ryan, D. H., Eds., Louisiana State University Press, Baton Rouge, 1992, 210.
21. **Mutt, V. and Jorpes, J. E.,** Hormonal polypeptides of the upper intestine, *Biochem. J.,* 125, 57, 1971.
22. **Iadarola, M. J., Narango, J. R., Duchemin, A. M., and Quach, T. T.,** Expression of cholecystokinin and enkephalin mRNA in discrete brain regions, *Peptides,* 10, 687, 1989.
23. **Rehfeld, J. F.,** Immunochemical studies on cholecystokinin. II. Distribution and molecular heterogeneity in the central nervous system in man and hog, *J. Biol. Chem.,* 253, 4022, 1978.
24. **Calam, J., Ellis, A., and Dockray, G. J.,** Identification and measurement of molecular variants of cholescystokinin in duodenal mucosa and plasma, *J. Clin. Invest.,* 69, 218, 1982.
25. **Smith, G. P., Jerome, C., Cushin, B. J., Eterno, R., and Simansky, K. J.,** Abdominal vagotomy blocks the satiety effect of cholecystokinin in the rat, *Science,* 213, 1036, 1981.
26. **Crawley, J. N. and Schwaber, J. S.,** Abolition of the behavioral effects of cholecystokinin following bilateral radiofrequency lesions of the parvocellular subdivision of the nucleus tractus solitarius, *Brain Res.,* 295, 289, 1984
27. **Edwards, G. L., Ladenheim, E. E., and Ritter, R. C.,** Dorsomedial hindbrain participation in cholecystokinin-induced satiety, *Am. J. Physiol.,* 251, R971, 1986.
28. **Crawley, J. N. and Kiss, J. Z.,** Paraventricular nucleus lesions abolish the inhibition of feeding induced by systemic cholecystokinin, *Peptides,* 6, 927, 1985.
29. **Silver, A. J. and Morley, J. E.,** Role of CCK in regulation of food intake, *Prog. Neurobiol.,* 36, 23, 1991.
30. **Anastasi, A., Erspamer, V., and Bucci, M.,** Isolation and structure of bombesin and alytesin, 2 analogous active peptides from the skin of the European amphibians Bombina and Alytes, *Experientia,* 27, 166, 1971.
31. **Polak, J. M., Bloom, S. R., Hobbs, S., Solcia, E., and Pearse, A. G.,** Distribution of a bombesin-like peptide in human gastrointestinal tract, *Lancet,* 1, 1109, 1976.
32. **Panula, P., Yang, H. Y., and Costa, E.,** Neuronal location of the bombesin-like immunoreactivity in the central nervous system of the rat, *Regul. Pept.,* 4, 275, 1982.
33. **Gibbs, J., Fauser, D. J., Rowe, E. A., Rolls, B. J., Rolls, E. T., and Maddison, S. P.,** Bombesin suppresses feeding in rats, *Nature,* 282, 208, 1979.
34. **Morley, J. E., Levine, A. S., Murray, S. S., and Kneip, J.,** Peptidergic regulation of norepinephrine induced feeding, *Pharmacol. Biochem. Behav.,* 16, 225, 1982.

35. **Morley, J. E. and Levine, A. S.**, Bombesin inhibits stress-induced eating, *Pharmacol. Biochem. Behav.*, 4, 149, 1981.
36. **Smith, G. P., Gibbs, J., Jerome, C., Pi-Sunyer, F. X., Kissileff, H. R., and Thornton, J.**, The satiety effect of cholecystokinin: a progress report, *Peptides*, 2 (Suppl. 2), 57, 1981.
37. **Stuckey, J. A. and Gibbs, J.**, Lateral hypothalamic injection of bombesin decreases food intake in rats, *Brain Res. Bull.*, 8, 617, 1982.
38. **Flynn, F. W.**, Fourth ventricular injection of selective bombesin receptor antagonists facilitates feeding in rats, *Am. J. Physiol.*, 264, R218, 1993.
39. **Merali, Z., Moody, T. W., and Coy, D.**, Blockade of brain bombesin/GRP receptors increases food intake in satiated rats, *Am. J. Physiol.*, 264, R1031, 1993.
40. **McCoy, J. G. and Avery, D. D.**, Bombesin: potential integrative peptide for feeding and satiety, *Peptides*, 11, 595, 1990.
41. **Carraway, R. and Leeman, S. E.**, The isolation of a new hypotensive peptide, neurotensin, from bovine hypothalami, *J. Biol. Chem.*, 248, 6854, 1973.
42. **Carraway, R. and Leeman, S. E.**, Characterization of radioimmunoassayable neurotensin in the rat: its differential distribution in the central nervous system, small intestine and stomach, *J. Biol. Chem.*, 251, 7045, 1976.
43. **Kobayashi, R. M., Brown, M. R., and Vale, W.**, Regional distribution of neurotensin and somatostatin in rat brain, *Brain Res.*, 126, 584, 1977.
44. **Levine, A. S., Kneip, J., Grace, M., and Morely, J. E.**, Effect of centrally administered neurotensin on multiple feeding paradigms, *Pharmacol. Biochem. Behav.*, 18, 19, 1983.
45. **Hawkins, M. F.**, Central nervous system neurotensin and feeding, *Physiol. Behav.*, 36, 1, 1986.
46. **Hoebel, B. G., Hernandez, L., McLean, S., Stanley, B. G., Aulissi, E. F., Glimcher, P., and Margolin D.**, in *The Neural Basis of Feeding and Reward*, Hoebel, B. G. and Novin, D., Eds., Haer Institute, Brunswick, 1982, 465.
47. **Hatfield, J. S., Millard, J., and Smith, C. J. V.**, Short-term influence of intra-ventromedial hypothalamic administration of insulin on feeding in normal and diabetic rats, *Pharmacol. Biochem. Behav.*, 2, 223, 1974.
48. **Havrankova, J., Roth, J., and Brownstein, M.**, Insulin receptors are widely distributed in the central nervous system of the rat, *Nature*, 272, 827, 1978.
49. **van Houten, M., Posner, B. I., Kopriwa, B. M., and Brawer, J. R.**, Insulin-binding sites in the rat brain: in vivo localization to the circumventricular organs by quantitative radioautography, *Endocrinology*, 105, 666, 1979.
50. **Margolis, R. U. and Altszulter, N.**, Insulin in the cerebrospinal fluid, *Nature*, 215, 1375, 1967.
51. **Woods, S. C. and Porte, D. Jr.**, Relationship between plasma and cerebrospinal fluid insulin levels of dogs, *Am. J. Physiol.*, 233, E331, 1977.
52. **Woods, S. C., McKay, D., Stein, L. J., and West, D. B.**, Insulin in the cerebrospinal fluid (CSF) of the baboon: response to feeding and intravenous (iv) infusions, *Diabetes*, 30, *(Abstr.)*, 119A, 1981.
53. **Strubbe, J. H. and Mein, C. G.**, Increased feeding in response to bilateral injection of insulin antibodies in the VMH, *Physiol. Behav.*, 19, 309, 1977.
54. **Woods, S. C., Lotter, E. C., McKay, L. D ., and Porte, D. Jr.**, Chronic intracerebroventricular infusion of insulin reduces food intake and body weight of baboons, *Nature*, 282, 503, 1979.
55. **Vanderweele, D. A., Pi-Sunyer, F. X., Novin, D., and Buch, M. J.**, Chronic insulin infusion suppresses food ingestion and body weight gain in rats, *Brain Res. Bull.*, 5 (Suppl. 4), 7, 1980.
56. **MacKay, E. M., Callaway, J. W., and Barnes, R. H.**, Hyperalimentation in normal animals produced by protamine insulin, *J. Nutr.*, 20, 59, 1940.
57. **Schulman, J. L., Carleton, J. L., Whitney, G., and Whitehorn, J. C.**, Effect of glucagon on food intake and body weight in man, *J. Appl. Physiol.*, 11, 419, 1957.

58. **Salter, J. M.,** Metabolic effects of glucagon in the Wistar rat, *Am. J. Clin. Nutr.*, 8, 535, 1960.
59. **Martin, J. R. and Novin, D.,** Decreased feeding in rats following hepatic-portal infusion of glucagon, *Physiol. Behav.*, 19, 461, 1977.
60. **Geary, N., Le-Sauter, J., and Noh, U.,** Glucagon acts in the liver to control spontaneous meal size in rats, *Am. J. Physiol.*, 264, R116, 1993.
61. **Geary, N. and Smith, G. P.,** Selective hepatic vagotomy blocks pancreatic glucagon's satiety effect, *Physiol. Behav.*, 31, 391, 1983.
62. **Langhans, W., Zeiger, U., Scharrer, E., and Geary, N.,** Stimulation of feeding in rats by intraperitoneal injection of antibodies to glucagon, *Science*, 218, 894, 1982.
63. **Nagai, K., Thibault, L., Nishikawa, K., Hashida, A., Ootani, K., and Nakagawa, H.,** Effect of glucagon in macronutrient self-selection: glucagon-enhanced protein intake, *Brain Res. Bull.*, 27, 409, 1991.
64. **Erlanson-Albertsson, C. and Larsson, A.,** The activation peptide of pancreatic procolipase decreases food intake in rats, *Reg. Pept.*, 22, 325, 1988.
65. **Okada, S., York, D. A., Bray, G. A., and Erlanson-Albertsson, C.,** Enterostatin, (Val-Pro-Asp-Pro-Arg), the activation peptide of procolipase selectivity reduces fat intake, *Physiol. Behav.*, 49, 1185, 1991.
66. **Erlanson-Albertsson, C. and Larsson, A.,** A possible physiological function of pancreatic pro-colipase activation peptide in appetite regulation, *Biochimie*, 70, 1245, 1988.
67. **Shargill, N. S., Tsujii, S., Bray, G. A., and Erlanson-Albertsson, C.,** Enterostatin suppresses food intake following injection into the third ventricle of rats, *Brain Res.*, 544, 137, 1991.
68. **Lin, L., McClanahan, S., York D. A., and Bray, G. A.,** The peptide enterostatin produces early satiety, *Physiol. Behav.*, 53, 789, 1993.
69. **Vale, W., Spiess, J., Rivier, C., and Rivier, J.,** Characterization of a 41-residue ovine hypothalamic peptide that stimulates secretion of corticotropin and B-endorphin, *Science*, 213, 1394, 1981.
70. **Britton, D. R., Koob, G. F., River, J., and Vale, W.,** Intraventricular corticotropin-releasing hormone enhances behavioral effects of novelty, *Life Sci.*, 31, 363, 1982.
71. **Morley, J. E. and Levine, A. S.,** Corticotropin-releasing factor, grooming and ingestive behavior, *Life Sci.*, 31, 1459, 1982.
72. **Krahn, D. D., Gosnell, B. A., Grace, M., and Levine, A. S.,** CRF antagonist partially reverses CRF- and stress-induced effects on feeding, *Brain Res. Bull.*, 17, 285, 1986.
73. **Gosnell, B. A., Morley, J. E., and Levine, A. S.,** A comparison of the effects of corticotropin releasing factor and sauvagine on food intake, *Pharmacol. Biochem. Behav.*, 19, 771, 1983.
74. **Krahn, D. D., Gosnell, B. A., Levine, A. S., and Morley, J. E.,** Behavioral effects of corticotropin-releasing factor: localization and characterization of central effects, *Brain Res.*, 443, 63, 1988.
75. **Glowa, J. R., Barret, J. E., Russell, J., and Gold, P. W.,** Effects of corticotropin releasing hormone on appetitive behaviors, *Peptides*, 13, 609, 1992.
76. **Guillemin, R., Yamazaki, E., Gard, D. A., Jutisz, M., and Sakiz, E.,** In vitro secretion of thyrotropin (TSH): stimulation by a hypothalamic peptide (TRF), *Endocrinology*, 73, 564, 1963.
77. **Morley, J. E., Garvin, T. J., Pekary, A. E., and Herschman, J. M.,** Thyrotropin-releasing hormone in the gatrointestinal tract, *Biochem. Biophys. Res. Commmun.*, 79, 314, 1977.
78. **Vijayan, E. and McCann, S. M.,** Suppression of feeding and drinking activity in rats following intraventricular injection of thyrotropin releasing hormone (TRH), *Endocrinology*, 100, 1727, 1977.
79. **Vogel, R. A., Cooper, B. R., Barlow, T. S., Prange, A. J. Jr., Mueller, R. A., and Breese, G. R.,** Effects of thyrotropin-releasing hormone on locomotor activity, operant performance and ingestive behavior, *J. Pharmacol. Exp. Ther.*, 208, 161, 1979.
80. **Morley, J. E., Levine, A. S., Kneip, J., and Grace, M.,** The effect of vagotomy on the satiety effects of neuropeptides and naloxone, *Life Sci.*, 30, 1943, 1982.

81. **Morley, J. E. and Levine, A. S.,** Thyrotropin releasing hormone (TRH) suppresses stress induced eating, *Life Sci.,* 27, 269, 1980.
82. **Prasad, C. and Peterkofsky, A.,** Demonstration of pyroglutamylpeptidase and amidase activities toward thyrotropin-releasing hormone in hamster hypothalamus, *J. Biol. Chem.,* 251, 3229, 1976.
83. **Morley, J. E., Levine, A. S., and Prasad, C.,** Histidyl-proline diketopiperazine decreases food intake in rats, *Brain Res.,* 210, 475, 1981.
84. **Clark, J. T., Kalra, P. S., Crowley, W. R., and Kalra, S. P.,** Neuropeptide Y and human pancreatic polypeptide stimulate feeding behavior in rats, *Endocrinology,* 115, 427, 1984.
85. **Levine, A. S. and Morley, J. E.,** Neuropeptide Y: a potent inducer of consummatory behavior in rats, *Peptides,* 5, 1025, 1984.
86. **Clark, J. T., Kalra, P. S., and Kalra, S. P.,** Neuropeptide Y stimulates feeding but inhibits sexual behavior in rats, *Endocrinology,* 117, 2435, 1985.
87. **Stanley, B. G. and Leibowitz, S. F.,** Neuropeptide Y: stimulation of feeding and drinking by injection into the paraventricular nucleus, *Life Sci.,* 35, 2635, 1984.
88. **Stanley, B. G., Kyrkouli, S. E., Lampert, S., and Leibowitz, S. F.,** Neuropeptide Y chronically injected into the hypothalamus: a powerful neurochemical inducer of hyperphagia and obesity, *Peptides,* 7, 1189, 1986.
89. **Stanley, B. G. and Leibowitz, S. F.,** Neuropeptide Y injected in the paraventricular hypothalamus: a powerful stimulant of feeding behavior, *Proc. Natl. Acad. Sci. U.S.A.,* 82, 3940, 1985.
90. **Chronwall, B. M., DiMaggio, D. A., Massari, V. J., Pickel, V. M., Ruggiero, D. A., and O'Donohue, T. L.,** The anatomy of neuropeptide Y-containing neurons in the rat brain, *Neuroscience,* 15, 1159, 1985.
91. **Stanley, B. G., Magdalin, W., Seirafi, A., Thomas, W. J., and Leibowitz, S. F.,** The perifornical area: the major focus of (a) patchily distributed hypothalamic neuropeptide Y-sensitive feeding system(s), *Brain Res.,* 604, 304, 1993.
92. **Jolicoeur, F. B., Bouali, S. M., Fournier, A., and St-Pierre, S.,** Mapping of hypothalamic sites involved in the effects of NPY on body temperature and food intake, *Brain Res. Bull.,* 36, 125, 1995.
93. **Stanley, B. G., Kyrkouli, S. E., Lampert, S., and Leibowitz, S. F.,** Neuropeptide Y chronically injected into the hypothalamus: a powerful neurochemical inducer of hyperphagia and obesity, *Peptides,* 7, 1189, 1986.
94. **Paez, X. and Myers, R. D.,** Insatiable feeding evoked in rats by recurrent perfusion of neuropeptide Y in the hypothalamus, *Peptides,* 12, 609, 1991.
95. **Zarjevski, N., Cusin, I., Vettor, R., Rohner-Jeanrenaud, F., and Jeanrenaud, B.,** Chronic intracerebroventricular neuropeptide-Y administration to normal rats mimics hormonal and metabolic changes of obesity, *Endocrinology,* 133, 1753, 1993.
96. **Beck, B., Burlet, A., Nicolas, J. P., and Burlet, C.,** Hypothalamic neuropeptide Y (NPY) in obese Zucker rats: implications in feeding and sexual behaviors, *Physiol. Behav.,* 47, 449, 1990.
97. **McKibbin, P. E., Cotton, S. J., McMillan, S., Holloway, B., Mayers, R., McCarthy, D., and Williams, G.,** Altered neuropeptide Y concentrations in specific hypothalamic regions of obese (fa/fa) Zucker rats, *Diabetes,* 40, 1423, 1991.
98. **Tatemoto, K., Rokaeus, A., Jornvall, H., McDonald, T. J., and Mutt, V.,** Galanin-a novel biologically active peptide from porcine intestine, *FEBS Lett.,* 164, 124, 1983.
99. **Kyrkouli, S. E., Stanley, B. G., and Leibowitz, S. F.,** Galanin: stimulation of feeding induced by medial hypothalamic injection of this novel peptide, *Eur. J. Pharm.,* 122, 159, 1986.
100. **Kyrkouli, S. E., Stanley, B. G., Seirafi, R. D., and Liebowitz, S. F.,** Stimulation of feeding by galanin: anatomical localization and behavioral specificity of the peptide's effects in the brain, *Peptides,* 11, 995, 1990.
101. **Schick, R. R., Samsami, S., Zimmermann, J. P., Eberl, T., Endres, C., Schusdziarra, V., and Classen, M.,** Effect of galanin on food intake in rats: involvement of lateral and ventromedial hypothalamic sites, *Am. J. Physiol.,* 264, R355, 1993.

102. **Leibowitz, S. F.,** Hypothalamic neuropeptide Y, galanin, amines: concepts of coexistence in relation to feeding behavior, *Ann. N.Y. Acad. Sci.,* 575, 221, 1989.
103. **Levin, M. C., Sawchenko, P. E., Howe, P. R. C., Bloom, S. R., and Polack, J. M.,** Organization of galanin-immunoreactive inputs to the paraventricular nucleus with special reference to their relationship to catecholaminergic afferents, *J. Comp. Neurol.,* 261, 562, 1987.
104. **Gundlach, A. L., Wisden, W., Morris, B. J., and Hunt, S. P.,** Localization of preprogalanin mRNA in rat brain: in situ hybridization study with a synthetic oligonucleotide probe, *Neurosci. Lett.,* 114, 241, 1990.
105. **Guillemin, R., Brazeau, P., Bohlen, P., Esch, F., Ling, N., and Wehrenberg, W. B.,** Growth hormone-releasing factor from a human pancreatic tumor that caused acromegaly, *Science,* 218, 585, 1982.
106. **Merchenthaler, I., Vigh, S., Shally, A. V., and Petrusz, P.,** Immunohistochemical localization of growth hormone-releasing factor in the rat hypothalamus, *Endocrinology,* 114, 1082, 1984.
107. **Vaccarino, F. J., Bloom, F. E., Rivier, J., Vale, W., and Koob, G. F.,** Stimulation of food intake in rats by centrally administered hypothalamic growth-hormone-releasing factor, *Nature,* 314, 167, 1985.
108. **Vaccarino, F. J, Feifel, D., Rivier, J., Vale, W., and Koob, G. F.,** Centrally administered rat hypothalamic growth hormone-releasing factor stimulates food intake in free-feeding rats, *Peptides,* 9 (Suppl. 1), 35, 1988.
109. **Imaki, T., Tamotsu, S., Hotta, M., Masuda, A., Demura, H., Shizumi, K., and Ling, N.,** The satiety effect of growth hormone-releasing factor, *Brain Res.,* 340, 186, 1985.
110. **Feifel, D. and Vaccarino, F. J.,** Feeding effects of growth hormone-releasing factor in rats are photoperiod sensitive, *Behav. Neurosci.,* 103, 824, 1989.
111. **Vaccarino, F. J. and Hayward, M.,** Microinjections of growth hormone-releasing factor into the medial preoptic area/suprachiasmatic nucleus region of the hypothalamus stimulate food intake in rats, *Reg. Peptides,* 21, 21, 1988.
112. **Vaccarino, F. J., Feifel, D., Rivier, J., and Vale, W.,** Antagonism of central growth hormone-releasing factor activity selectively attenuates dark-onset feeding in rats, *J. Neurosci.,* 11, 3924, 1991.
113. **Barbour, H. G., Gregg, D. E., and Hunter, L. G.,** The calorigenic action of morphine as revealed by addiction studies, *J. Pharmacol. Exp. Ther.,* 40, 433, 1930.
114. **Holtzman, S. G.,** Behavioral effects of separate and combined administration of naloxone and d-amphetamine, *J. Pharmacol. Exp. Ther.,* 189, 51, 1974.
115. **Khachaturian, H., Lewis, M. E., Schafer, M. K.-H., and Watson, S. J.,** Anatomy of the CNS opioid systems, *Trends Neuro. Sci.,* 8, 111, 1985.
116. **Walker, J. M., Datz, R. J., and Akil, H.,** Behavioral effect of dynorphin-1-13 in the mouse and rat: initial observations, *Peptides,* 1, 341, 1980.
117. **Morley, J. E. and Levine, A. S.,** Dynorphin-(1-13) induces spontaneous feeding in rats, *Life Sci.,* 29, 1901, 1981.
118. **Schulz, R., Wilhelm, A., and Dirlich, G.,** Intracerebral injection of different antibodies against endogenous opioids suggested alpha-neo-endorphin participation in control of feeding behavior, *Naunyn. Schmiedebergs Arch. Pharmacol.,* 326, 222, 1984
119. **Arjune, D. and Bodner, R. J.,** Suppression of nocturnal, palatable and glucoprivic intake in rats by the K opioid antagonist, nor-binaltorphamine, *Brain Res.,* 534, 313, 1990.
120. **Grandison, L. and Guidotti, A.,** Stimulation of food intake by muscimol and beta-endorphin, *Neuropharmacology,* 16, 533, 1977.
121. **McKay, L. D., Kenney, N. J., Edens, N. K., Williams, R. H., and Woods, S. C.,** Intracerebroventricular beta-endorphin increases food intake in rats, *Life Sci.,* 29, 1429, 1981.
122. **Levine, A. S., Morley, J. E., Gosnell, B. A., Billington, C. J., and Bartness, T. J.,** Opioids and consummatory behavior, *Brain Res. Bull.,* 14, 663, 1985.

123. **Sahu, A., Kalra, P. S., and Kalra, S. P.,** Food deprivation and ingestion induce reciprocal changes in neuropeptide Y concentrations in the paraventricular nucleus, *Peptides,* 9, 83, 1988.
124. **White, J. D. and Kershaw, M.,** Increased hypothalamic neuropeptide Y expression following food deprivation, *Mol. Cell. Neurosci.,* 1, 41, 1990.
125. **Sanacora, G., Kershaw, M., Finkelstein, J. A., and White, J. D.,** Increased hypothalamic content of preproneuropeptide Y messenger ribonucleic acid in genetically obese Zucker rats and its regulation by food deprivation, *Endocrinology,* 127, 730, 1990.
126. **Brady, L. S., Smith, M. A., Gold, P. W., and Herkenham, M.,** Altered expression of hypothalamic neuropeptide mRNAs in food-restricted and food-derived rats, *Neuroendocrinology,* 52, 441, 1990.
127. **O'Shea, R. D. and Gundlach, A. L.,** Preproneuropeptide Y messenger ribonucleic acid in the hypothalamic arcuate nucleus of the rat is increased by food deprivation or dehydration, *J. Neuroendocrinol.,* 3, 11, 1991.
128. **Bai, F. L., Yamano, M., Shiotani, Y., Emson, P., Smith, A. D., Powell, J. F., and Tohyama, M.,** An arcuato-paraventricular and -dorsomedial hypothalamic neuropeptide Y-containing system which lacks noradrenaline in the rat, *Brain Res.,* 331, 172, 1985.
129. **Stanley, B. G., Magdalin, W., Seifafi, A., Nguyen, M. M., and Leibowitz, S. F.,** Evidence for neuropeptide Y mediation of eating produced by food deprivation and for a variant of the Y1 receptor mediating this peptide's effect, *Peptides,* 13, 581, 1992.
130. **Schwartz, M. W., Sipols, A. J., Grubin, C. E., and Baskin, D. G.,** Differential effects of fasting on hypothalamic expression of genes encoding neuropeptide Y, galanin, and glutamic acid decarboxylase, *Brain Res. Bull.,* 31, 361, 1993.
131. **Bruno, J. F., Olchovsky, D., White, J. D., Leidy, J. W., Song, J., and Berelowitz, M.,** Influence of food deprivation in the rat on hypothalamic expression of growth hormone-releasing factor and somatostatin, *Endocrinology,* 127, 2111, 1990.
132. **Merali, Z. and Kateb, C. C.,** Rapid alterations of hypothalamic and hippocampal bombesin-like peptide levels with feeding status, *Am. J. Physiol.,* 265, R420, 1993.
133. **Siegel, P. S.,** Food intake in the rat in relation to the light–dark cycle, *J. Comp. Physiol. Psychol.,* 54, 294, 1961.
134. **Jhanwar-Uniyal, M., Beck, B., Burlet, C., and Leibowitz, S. F.,** Diurnal rhythm of neuropeptide Y-like immunoreactivity in the suprachiasmatic, arcuate and paraventricular nuclei and hypothalamic sites, *Brain Res.,* 536, 331, 1990.
135. **Akabayashi, A., Levin, N., Paez, X., Alexander, J. T., and Leibowitz, S. F.,** Hypothalamic neuropeptide Y and its gene expression: relation to light/dark cycle and circulating corticosterone, *Mol. Cell. Neurosci.,* 5, 210, 1994.
136. **White, B. D., Dean, R. G., and Martin, R. J.,** Short-term dexamethasone treatment increases food intake, but apparently not through changes in NPY content in the PVN, *Ann. NY Acad. Sci.,* 739, *(Abstr.),* 339, 1994.
137. **Kalra, S. P., Dube, M. G., Sahu, A., Phelps, C., and Kalra, P. S.,** Neuropeptide Y secretion increases in the paraventricular nucleus in association with increased appetite for food, *Proc. Natl. Acad. Sci. U.S.A.,* 88, 10931, 1991.
138. **Krieger, D. T.,** Food and water restriction shifts corticosterone, temperature, activity and brain amine periodicity, *Endocrinology,* 95, 1195, 1974.
139. **Wahlestedt, C., Skagerberg, G., Edman, R., Heilig, M., Sundler, F., and Hakanson, R.,** Neuropeptide Y (NPY) in the area of the paraventricular nucleus activates the pituitary-adrenalecortical axis in the rat, *Brain Res.,* 417, 33, 1987.
140. **Leibowitz, S. F., Sladek, C., Spencer, L., and Tempel, D.,** Neuropeptide Y, epinephrine and norepinephrine in the paraventricular nucleus: stimulation of feeding and the release of corticosterone, vasopressin and glucose, *Brain Res. Bull.,* 21, 905, 1988.
141. **Corder, R., Pralong, F., Turnill, D., Saudan, P., Muller, A. F., and Gaillard, R. C.,** Dexamethasone treatment increases neuropeptide Y levels in rat hypothalamic neurons, *Life Sci.,* 43, 879, 1988.

142. **McKibbin, P. E., Cotton, S. J., McCarthy, H. D., and Williams, G.,** The effect of dexamethasone on neuropeptide Y concentrations in specific hypothalamic regions, *Life Sci.,* 51, 1301, 1992.
143. **White, B. D., Dean, R. G., and Martin, R. J.,** Adrenalectomy decreases neuropeptide Y mRNA levels in the arcuate nucleus, *Brain Res. Bull.,* 25, 711, 1990.
144. **Ponsalle, P., Srivastava, L. S., Uht, R. M., and White, J. D.,** Glucocorticoids are required for food deprivation-induced increases in hypothalamic neuropeptide Y expression, *J. Neuroendocrinol.,* 4, 585, 1992.
145. **White, B. D., Dean, R. J., Edwards, G. L., and Martin, R. J.,** Type II corticosteroid receptor stimulation increases NPY gene expression in basomedial hypothalamus of rats, *Am. J. Physiol.,* 266, R1523, 1994.
146. **Linden, A. and Uvnas-Maberg, K.,** Plasma levels of cholecystokinin (CCK-8 and CCK-33-39) in response to feeding and during pregnancy in dogs, *Scand. J. Gastroentero.,* 22, 859, 1987.
147. **Schick, R. R., Yaksh, T. L., Roddy, D. R., and Go, V. L.,** Release of hypothalamic cholecystokinin in cats: effects of nutrient and volume loading, *Am. J. Physiol.,* 256, R248, 1989.
148. **West, D. B., Fey, D., and Woods, S. C.,** Cholecystokinin persistently suppresses meal size but not food intake in free-feeding rats, *Am. J. Physiol.,* 246, R776, 1984.
149. **Mashford, M. L., Nilsson, G., Rokaeus, A., and Rosell, S.,** The effect of food ingestion on circulating neurotensin-like immunoreactivity (NTLI) in the human, *Acta Physiol. Scand.,* 104, 244, 1978.
150. **Shaw, C. and Buchanan, K. D.,** Intact neurotensin (NT) in human plasma: response to oral feeding, *Reg. Pept.,* 7, 145, 1983.
151. **George, J. K., Albers, H. E., Carraway, R. E., and Ferris, C. F.,** Neurotensin levels in the hepatic-portal circulation are inversely related to the circadian feeding cycle in rats, *Endocrinology,* 121, 7, 1987.
152. **Fujita, Y, Gotto, A. M., Phil, D., and Unger, R. H.,** Basal and postprotein insulin and glucagon levels during a high and low carbohydrate intake and their relationship to plasma triglycerides, *Diabetes,* 24, 552, 1975.
153. **Mei, J., Bowyer, R. C., Jehanli, A. M., Patel, G., and Erlanson-Albertsson, C.,** Identification of enterostatin, the pancreatic procolipase activation peptide in the intestine of rat: effect of CCK-8 and high fat feeding, *Pancreas,* 8, 488, 1993.
154. **Tempel. D. L., Shor-Posner, G., Dwyer, D., and Leibowitz, S. F.,** Nocturnal patterns of macronutrient intake in freely feeding and food-deprived rats, *Am. J. Physiol.,* 256, R541, 1989.
155. **Stanley, B. G., Daniel, D. R., Chin, A. S., and Leibowitz, S. F.,** Paraventricular nucleus injections of peptide YY and neuropeptide Y preferentially enhance carbohydrate ingestion, *Peptides,* 6, 1205, 1985.
156. **Jhanwar-Uniyal, M., Beck, B., Jhanwar, Y. S., Burlet, C., and Leibowitz, S. F.,** Neuropeptide Y projection from arcuate nucleus to parvocellular division of paraventricular nucleus: specific relation to the ingestion of carbohydrate, *Brain Res.,* 631, 97, 1993.
157. **Akabayashi, A., Wahlestedt, C., Alexander, J. T., and Leibowitz, S. F.,** Specific inhibition of endogenous neuropeptide Y synthesis in arcuate nucleus by antisense oligonucleotides suppresses feeding behavior and insulin secretion, *Brain Res.,* 21, 55, 1994.
158. **Stanley, B. G., Anderson, K. C., Grayson, M. H., and Leibowitz, S. F.,** Repeated hypothalamic stimulation with neuropeptide Y increases daily carbohydrate and fat intake and body weight gain in female rats, *Physiol. Behav.,* 46, 173, 1989.
159. **Welch, C. C., Grace, M. K., Billington, C. J., and Levine, A. S.,** NPY induces fat selection in fat-preferring rats, *FASEB J.,* 7, *(Abstr.),* A88, 1993.
160. **Welch, C. C., Grace, M. K., Billinton, C. J., and Levine, A. S.,** Preference and diet type affect macronutrient selection after morphine, NPY, norepinephrine and deprivation, *Am. J. Physiol.,* 266, R426, 1994.
161. **Tempel, D. L. and Leibowitz, S. F.,** Diurnal variations in the feeding responses to norepinephrine, neuropeptide Y and galanin in the PVN, *Brain Res. Bull.,* 25, 821, 1990.

162. **Leibowitz, S. F. and Kim, T.,** Impact of a galanin antagonist on exogenous galanin and natural patterns of fat ingestion, *Brain Res.*, 599, 148, 1992.
163. **Marks-Kaufman, R. and Kanarek, R. B.,** Morphine selectively influences macronutrient intake in the rat, *Pharmacol. Biochem. Behav.*, 12, 427, 1980.
164. **Marks-Kaufman, R. and Kanarek, R. B.,** Modifications of nutrient selection induced by naloxone in rats, *Psychopharmacology,* 74, 321, 1981.
165. **Romsos, D. R., Gosnell, B. A., Morley, J. E., and Levine, A. S.,** Effects of kappa opiate agonists, cholecystokinin and bombesin on intake of diets varying in carbohydrate-to-fat ratio in rats, *J. Nutr.*, 117, 976, 1987.
166. **Okada, S., Lin, L., York, D. A., and Bray, G. A.,** Chronic effects of intracerebral ventricular enterostatin in Osborne-Mendel rats fed a high fat diet, *Physiol. Behav.*, 54, 325, 1993.
167. **Mizuma, H., Abadie, J., and Prasad, C.,** Corticosterone facilitation of inhibition of fat intake by enterostatin (Val-Pro-Asp-Pro-Arg), *Peptides,* 15, 447, 1994.
168. **Lin, L., York, D., and Bray, G.,** Acute effects of intracerebroventriclar corticotropin releasing hormone (CRH) on macronutrient selection, *Int. J. Obes.*, 16, *(Abstr.),* P207, 1992.
169. **Dickson, P. R. and Vaccarino, F. J.,** GRF-induced feeding: evidence for protein selectivity and opiate involvement, *Peptides,* in press.
170. **Dickson, P. R., Feifel, D., and Vaccarino, F. J.,** Blockade of endogenous GRF at dark onset selectively suppresses protein intake, *Peptides,* 16, 7, 1995.
171. **Avery, D. D. and Livosky, M.,** Peripheral injections of bombesin and cholecystokinin affect dietary self-selection in rats, *Pharmacol. Biochem. Behav.*, 25, 7, 1986.
172. **Thibault, L., Nagai, K., Hashida, A., Yanaihara, N., and Nakagawa, H.,** Satiation due to CCK-8 derivation infusion into VMH is related to a specific macronutrient selection, *Physiol. Behav.*, 47, 911, 1990.
173. **Bruno, J. F., Song, J., and Berelowitz, M.,** Regulation of rat hypothalamic preprogrowth hormone-releasing factor messenger ribonucleic acid by dietary protein, *Endocrinology,* 129, 1226, 1991.
174. **Kanarek, R. B., Marks-Kaufman, R., Ruthazer, R., and Gualtieri, L.,** Increased carbohydrate consumption by rats as a function of 2-deoxy-D-glucose administration, *Pharmacol. Biochem. Behav.*, 18, 47, 1983.
175. **Akabayashi, A., Thais, C., Zaia, B. V., Silva, I., Chae, H. J., and Leibowitz, S. F.,** Neuropeptide Y in the arcuate nucleus is modulated by alterations in glucose utilization, *Brain Res.*, 621, 343, 1993.
176. **McDonald, J. K.,** NPY and related substances, *Crit. Rev. Neuro.*, 4, 97, 1988.
177. **Crowley, W. R., Shah, G. V., Carroll, B. L., Kennedy, D., Dockter, M. E., and Kalra, S. P.,** NPY enhances leutinizing hormone (LH)-releasing hormone-induced LH release and elevations in cytosolic Ca^{++} in rat anterior pituitary cells: evidence for involvement of extracellular Ca^{++} influx through voltage sensitive channels, *Endocrinology,* 127,1487, 1990.
178. **Ashendel, C. L.,** The phorbol ester receptor: a phospholipid regulated protein kinase, *BBA,* 822, 219, 1985.
179. **Nishizua, Y.,** Studies and perspectives of protein kinase C, *Science,* 233, 305, 1986.
180. **Seamon, K. D., Padgett, W., and Daly, J. W.,** Forskolin: unique diterpene activator of adenylate cyclase in membranes and in intact cells, *Proc. Natl. Acad. Sci. U.S.A.*, 78, 3363, 1981.
181. **Morgan, J. I. and Curran, T.,** Stimulus-transcription coupling in the nervous system: involvement of the inducible proto-oncogenes fos and jun, *Ann. Rev. Neurosci.*, 14, 421, 1991.
182. **Higuchi, H., Yang, H-Y. T., and Sabol, S. L.,** Rat neuropeptide Y precursor gene expression, *J. Biol. Chem.*, 263, 6288, 1988.
183. **Sabol, S. and Higuchi, H.,** Transcriptional regulation of NPY gene by nerve growth factor: antagonism by glucocorticoids and potentiation by adenosine 3¢ 5¢ monophosphate and phorbol ester, *Mol. Endocrinol.*, 4, 384, 1990.

184. **Minth-Worby, C. A.,** Transcriptional regulation of human NPY gene by nerve growth factor, *J. Biol. Chem.,* 269, 15460, 1994.
185. **Minth, C. D. and Dixon, J. E.,** Expression of human NPY gene, *J. Biol. Chem.,* 256, 12933, 1990.
186. **Lerchen, R. A., Yum, D-Y., Krajcik, R., and Minth-Worby, C. A.,** Transcriptional vs. posttranscriptional control of NPY gene expression, *Endocrinology,* 136, 833, 1995.
187. **He, B., White, B. D., and Martin, R. J.,** Neuropeptide Y does not appear to be involved in 2-deoxyglucose induced feeding, *FASEB J.,* 7, *(Abstr.),* A88, 1993.
188. **Giraudo, S. Q., Grace, M. K., Levine, A. S., and Billington, C. J.,** Central and peripheral 2-DG: effect on neuropeptide Y gene expression and peptide level, *Soc. Neurosci.,* 20, *(Abstr.),* 1223, 1994.
189. **Smith, M. S.,** Lactation alters neuropeptide-Y and proopiomeleanocortin gene expression in the arcuate nucleus of the rat, *Endocrinology,* 133, 1258, 1993.
190. **Beck, B., Stricker-Krongrad, A., Burlet, A., Nicolas, J-P., and Burlet, C.,** Influence of diet composition on food intake and hypothalamic neuropeptide Y (NPY) in the rat, *Neuropeptides,* 17, 197, 1990.
191. **Beck, B., Sticker-Krongrad, A., Burlet, A., Max, J-P., Musse, N., Nicolas, J-P., and Burlet, C.,** Macronutrient type independently of energy intake modulates hypothalamic neuropeptide Y in Long-Evans rats, *Brain Res. Bull.,* 34, 85, 1994.
192. **Sclafani, A. and Springer, D.,** Dietary obesity in adult rats: similarities to hypothalamic and human obesity syndromes, *Physiol. Behav.,* 17, 461, 1976.
193. **Wilding, J. P. H., Gilbey, S. G., Mannan, M., Aslam, N., Ghatei, M. A., and Bloom, S. R.,** Increased neuropeptide Y content in individual hypothalamic nuclei, but not neuropeptide Y mRNA, in diet-induced obesity in rats, *J. Endocrinol.,* 132, 299, 1992.
194. **Larson, P. J., Mikkelsen, J. D., Jessop, D. S., Lightman, S. L., and Chowdrey, H. S.,** Neuropeptide Y mRNA and immunoreactivity in hypothalamic neuroendocrine neurons: effect of adrenalectomy and chronic osmotic stimulation, *J. Neurosci.,* 13, 1138, 1993.
195. **Giraudo, S. Q., Kotz, C. M., Grace, M. K., Levine, A. S., and Billington, C. J.,** Rat hypothalamic NPY mRNA and brown fat uncoupling protein mRNA after high carbohydrate or high fat diets, *Am. J. Physiol.,* 266, R1578, 1994.
196. **White, B. D., He, B., Dean, R. G., and Martin, R. J.,** Low protein diets increase neuropeptide Y gene expression in the basomedial hypothalamus of rats, *J. Nutr.,* 124, 1152, 1994.
197. **Kuczmarski, R. J., Flegal, K. M., Campbell, S. M., and Johnson, C. L.,** Increasing prevalence of overweight among US adults, *JAMA,* 272, 205, 1994.

Chapter 5

GENETIC AND ENERGY BALANCE INTERACTIONS IN HUMANS

Claude Bouchard

CONTENTS

INTRODUCTION

Most research reported to date on the genetic basis of human obesity or energy balance determinants has been conducted with the tools of genetic epidemiology. These studies have been helpful in understanding the role of inherited differences in a population perspective, but they are not able to clarify issues such as the role of genetic variation in response to dietary treatments or various dietary practices. To deal with these topics, one must use a variety of experimental and molecular methods that have not been used extensively so far in humans. We have reached the stage where it has become possible to

0-8493-9485-6/96/$0.00+$.50

contemplate "gene-directed nutrient intake recommendations"[1] as a realistic goal for the not too distant future. To meet this goal, it will be important to focus some of our research efforts on the role of biological individuality in caloric and nutrient needs. It is well established that there are considerable individual differences in response to various dietary manipulations such as a diet rich in cholesterol, or rich in saturated fats. The same phenomenon has been observed in response to chronic overfeeding. For instance, the Vermont overfeeding study showed that body weight gain varied among subjects and that it was below the expected weight gain based upon the ingested caloric surplus.[2] Much has been learned from experimental overfeeding, dietary energy restriction or exercise-induced negative energy balance studies conducted with human subjects.

We have proposed that such experiments would yield better and more powerful data if they were performed with sets of monozygotic (MZ) twins.[3] We have elected to use this model in our experimental genetic studies of the causes and outcomes of variations in body fat and regional fat distribution and of variations in response to chronic alterations in energy balance. A unique aim of this research is to try to determine whether there is a genotype–environment interaction effect in the response to the treatment. Environment is defined here as the overfeeding or negative energy balance treatment. A genotype–environment interaction effect refers to a phenotype for which the response to an environmental challenge is significantly influenced by the genotype.[4,5] To date, we have completed four of these studies, two overfeeding experiments and two negative energy balance studies. These studies were conducted with the collaboration of a total of 31 pairs of male MZ twins, ranging in age from 17 to 29 years. The main results of these four studies will be reviewed here.

THE EXPERIMENTAL DESIGN

It is important to understand the essential features of the experimental MZ design. The procedure requires that both members of several pairs of MZ twins be subjected exactly to the same experimental treatment.[3] If there are individual differences in response to the treatment, within-pairs and between-pairs variances can be obtained. Full control over all aspects of the treatment is critical to the success of such an experiment.

Some of the conditions that are important to consider can be summarized as follows: (a) determine twin zygosity as precisely as possible; (b) keep age variation at a minimum; (c) use twin pairs of only one gender or control for gender difference if male and female twin pairs are involved; (d) apply the treatment in exactly the same manner to all twins under rigorously controlled conditions; and (e) select phenotypes that are not, or are little affected, by prior exposure to the treatment variable of interest in the study.[3] The procedure is quite similar to that used in animal genetics when testing for a genotype–environment interaction effect by comparing various strains of a given species

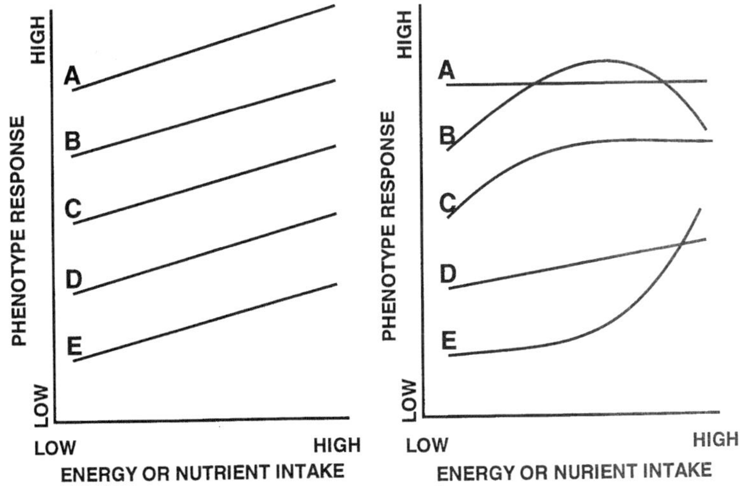

FIGURE 1. Phenotype response with chronic exposure to different levels of energy or nutrient intake conditions. In the left panel, the changes in the phenotype are not influenced by the genotype (A = B = C = D = E responses). In the right panel, the genotype determines the response to the dietary treatment, i.e., there is a genotype–diet interaction effect (A # B # C # D # E responses).

exposed to a given treatment. The main difference is that by using MZ twins, we have only two subjects per genotype. However, the number of genotypes (MZ pairs) can be relatively large depending on the requirements of the research.

Data gathered before and after treatment on both members of several twin pairs can be analyzed by a two-way analysis of variance for repeated measures on one factor (the treatment effect). Twins are nested in pairs. F ratios for the treatment effect and for the genotype–treatment interaction effect can be obtained. The intraclass correlation computed with the absolute or relative response to the treatment can also be computed from the within- and between-pairs means of squares, without or with adjustment for the pre-treatment level.[3]

A larger between-MZ-pairs variance compared to the within-pair variance suggests that the response to the experimental treatment is more heterogeneous in subjects who are genetically different. If a pair of MZ twins is assumed to represent a given genotype (for all relevant loci), then one can illustrate the within- and between-genotype difference under the scenario of no genotype–diet interaction effect (left panel of Figure 1) and that with a significant genotype–diet interaction effect (right panel of Figure 1). In a series of experiments conducted with MZ twin pairs over the last 10 years in our laboratory, we have shown that the experimental model was useful in understanding the role of the genotype–treatment interactions for such variables as chronic overfeeding, regular exercise, and regular exercise with negative energy balance. In general, these studies have revealed that genotype–environment interaction effects were ubiquitous.

There are, however, important limitations with this design, particularly from the point of view of what can be undertaken experimentally and for ethical considerations. For instance, even though it is possible, at least from a theoretical point of view, to exert satisfactory experimental control and reach full standardization over energy or nutrient intake in overfeeding or underfeeding studies, it is not possible to fully standardize energy expenditure. Clamping of energy expenditure at the same level for all subjects of a given group cannot be achieved because of individual differences in resting metabolic rate, thermic response to food, fidgeting, or variations in body mass that impact on the energy cost of weight maintenance. Of course, there are limits as to the severity of the nutritional stress that can be tested as well as in the duration of the experimental treatment. Such studies are also financially very costly to undertake. Nonetheless, they provide us with unique data when performed with MZ twins.

The method can be extended to experimental studies with singletons and relevant molecular markers. Indeed, it is possible to investigate the response to chronic alterations in energy balance or nutrient intake in relation to genetic characteristics at given candidate genes and/or sets of molecular markers. This approach has the potential to identify the genes that are associated or linked (with siblings and dizygotic twins) with individual differences in the response to the treatment. For instance, such studies could yield important data on the genes responsible for the susceptibility to be in positive energy balance or gain weight over time, as well as those associated with the resistance to weight loss. The approach can be used to delineate the quantitative importance of the genotype–treatment interaction effects for a variety of complex multifactorial phenotypes or risk factors for common nutritional deficiencies and diseases, and to identify their molecular basis.

THE SHORT-TERM EXPERIMENTS

It is generally recognized that there are some individuals prone to excessive accumulation of fat, for which losing weight represents a continuous battle, and others who seem relatively well protected against such a menace. We have recently tried to test whether such differences could be accounted for by inherited differences. In other words, we asked whether there were differences in the sensitivity of individuals to gain fat when chronically exposed to positive energy balance or to lose fat when exposed to negative energy balance conditions and whether such differences were dependent or independent of the genotype. If the answer to these questions was affirmative then one would have to conclude that there was a significant genotype–energy balance interaction effect. The results from two short-term experiments (22 days) suggested that such an effect was likely to exist for body weight, body fat, fat distribution, metabolic rates, and several aspects of carbohydrate and lipid metabolism. These results were subsequently confirmed by long-term studies which will be reviewed in the next section.

In the first study, we exposed 6 pairs of male MZ twins to a 1000 kcal per day energy intake surplus for a period of 22 consecutive days.[6,7] Individual differences in body weight, fat mass, subcutaneous fat, and site of fat deposition gains were observed with this short-term overfeeding protocol, but these differences were not randomly distributed. Indeed, significant intrapair resemblance was observed for the changes in most body composition and fat distribution variables despite the fact that the treatment was of short duration and that the changes induced by the treatment were not large. The intrapair resemblance in the response to overfeeding as assessed by the intraclass coefficient computed with the individual changes, reached 0.88 for total fat mass and 0.76 for fat-free mass. Subjects gained body weight and body fat but there was a nonsignificant 7% increase in resting metabolic rate.[8] Interindividual differences in resting metabolic rate were, however, considerable but not randomly distributed as suggested by a significant F ratio for the between-pairs over within-pairs variances ($F = 4.45$; $p < 0.05$) for resting metabolic rate expressed in ml of O_2 per minute. The intraclass correlation computed to quantify the within-pair resemblance in the response reached 0.63, but it decreased to a nonsignificant level when changes in resting metabolic rate were adjusted for gains in body mass or fat-free mass.[7] A similar trend was observed for thermic effect of a meal (intraclass = 0.62) and energy cost of submaximal exercise (intraclass = 0.78).[7,9]

On the other hand, no major alterations in glucose and insulin response to a glucose load or a test meal, in cardio-pulmonary adaptation to submaximal exercise and in maximal exercise tolerance were found with overfeeding.[7] In contrast, the response of suprailiac fat cell lipolysis (intraclass of about 0.7) and heparin releasable adipose tissue lipoprotein lipase (intraclass = 0.82) varied among individuals but was highly homogeneous within genotypes.[10] Similarly, a genotype–overfeeding interaction effect was seen for serum triglycerides (intraclass = 0.69), HDL-cholesterol (intraclass = 0.85), and the HDL-cholesterol to total cholesterol ratio (intraclass = 0.82). The data suggested that the genetic effect observed in the response of HDL-cholesterol and the ratio of HDL-cholesterol to total cholesterol to overfeeding was associated with the susceptibility to develop hypertriglyceridemia when challenged by the caloric surplus. The changes in triglyceride levels were themselves well correlated with the changes in the fasting insulin and post-glucose insulin area.[11]

A short-term negative energy balance experiment was also undertaken with six pairs of MZ twins. The energy deficit was achieved by exercise performed twice a day, for 22 consecutive days, for about 50 min per session, on the cycle ergometer. The exercise prescription was precisely controlled for each subject, during each exercise session, and was designed to induce an extra energy expenditure of 1000 kcal/day over resting metabolic rate while baseline energy intake was maintained throughout.[12] The changes in body composition were generally small and not related to the twin lines. The only exception was for the fat free mass changes which were more similar within pairs in comparison to the between pairs-variance.

The program induced a significant increase in VO_2max and a significant decrease in body fat content. Significant changes were observed in fasting plasma insulin and in the insulin response to an oral glucose tolerance test.[13] Plasma triglycerides, total cholesterol, LDL-cholesterol, and apo B as well as the HDL-cholesterol to total cholesterol ratio were also modified.[14] Moreover, significant within-MZ-pair resemblance was observed for the response of fasting plasma insulin, LDL-cholesterol, HDL-cholesterol, and the HDL-cholesterol to total cholesterol ratio. Although this short-term exercise training program did not produce significant changes in resting metabolic rate and thermic response to food, there was heterogeneity in the response. Once again, a significant intrapair resemblance was observed for the changes in resting metabolic rate ($p < 0.05$) and thermic response to food ($p < 0.05$).[15]

THE LONG-TERM EXPERIMENTS

The short-term (22 days) experiments provided useful indications about the genotype–energy balance interaction effects, but they were based on experimental responses that were not large. We felt that more could be learned if larger changes could be induced so that the heterogeneity in response could be better delineated. To this end, we undertook two long-term intervention studies that will now be summarized.

LONG-TERM OVERFEEDING EXPERIMENT

Body Composition and Energy Balance

Twelve pairs of male MZ twins ate a 1000 kcal per day caloric surplus, 6 days a week, over a period of 100 days.[16] Significant increases in body weight and fat mass were observed after the period of overfeeding. Data showed that there were considerable interindividual differences in the adaptation to excess calories and that the variation observed was not randomly distributed, as indicated by the significant within-pair resemblance in response. For instance, there was at least three times more variance in response for between-pairs than for within-pairs for the gains in body weight, fat mass, and fat-free mass (Table 1 and Figure 2). These data, and those that we have reported earlier for the response to short-term overfeeding, demonstrate that some individuals are more at risk than others to gain fat when energy intake surplus is clamped at the same level for everyone and when all subjects are confined to a sedentary lifestyle. The within-pair identical twin response to the standardized caloric surplus suggests that the amount of fat stored is likely influenced by the genotype.

At the beginning of the overfeeding treatment, almost all the daily caloric surplus was recovered as body energy gain, but the proportion decreased to 60% at the end of the 100-day protocol.[17] The weight gain pattern followed an exponential with a half-duration of about 86 days. We have estimated that the

TABLE 1
EFFECTS OF THE 100-DAY OVERFEEDING TREATMENT AND INTRAPAIR RESEMBLANCE IN THE ABSOLUTE RESPONSE

Variable	Before Overfeeding (X ± SD)	After Overfeeding (X ± SD)[a]	Within-Pair Resemblance F Ratio[b]	Within-Pair Resemblance Intraclass
Body weight (kg)	60.3 ± 8.0	68.4 ± 8.2	3.43	0.55
Body mass index (kg/m)	19.7 ± 2.0	22.4 ± 2.0	2.85	0.48
Percent fat	11.3 ± 5.0	17.8 ± 5.7	2.92	0.49
Fat mass (kg)	6.9 ± 3.5	12.3 ± 4.5	3.00	0.50
Fat-free mass (kg)	53.4 ± 6.6	56.1 ± 6.7	2.34[c]	0.40[c]
Fat mass/fat-free mass	0.13 ± 0.06	0.22 ± 0.08	3.30	0.53
Subcutaneous fat (mm)	75.9 ± 21.2	129.4 ± 32.9	2.77	0.47
Body energy (MJ)	497 ± 142	719 ± 176	3.12	0.51

Note: Statistical significance was established from a two-way analysis of variance for repeated measures on one factor (time). F ratio of between-pairs over within-pairs variances. Intraclass coefficient is used to assess the within-pair resemblance in response to the treatment. (Reproduced from Ref. 16 with permission of the Massachusetts Medical Society.)

a $p < 0.0001$.
b $p < 0.05$.
c non-significant.

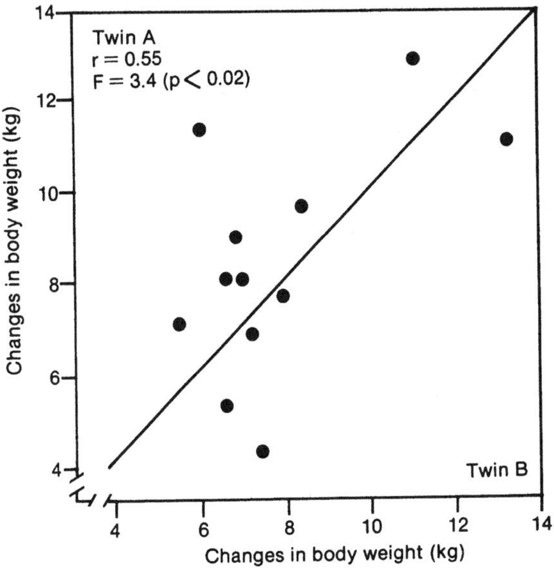

FIGURE 2. Changes in body weight in response to a 100-day overfeeding protocol in 12 pairs of young adult male identical twins. The within-pair resemblance in response is illustrated. (Reproduced from Ref. 16 by permission of the Massachusetts Medical Society.)

weight gain attained in the experiment reached about 55% of the anticipated maximal weight gain had the overfeeding protocol been continued indefinitely.[17]

The mean body mass gain for the 24 subjects of the 100-day overfeeding experiment was 8.1 kg, of which 5.4 kg were fat mass and 2.7 kg were fat-free mass increases. Assuming that the energy content of body fat is about 9300 kcal/kg and that of fat-free tissue is 1020 kcal/kg, then about 63% of the excess energy intake was recovered on the average as body mass changes. This proportion is on the same order as that reported by other investigators,[18,19] i.e., between 60 and 75% of total excess energy intake. There were, however, individual differences among the 24 subjects with respect to the amount of fat and fat-free tissues gained. Thus, while the mean gain of fat mass to fat-free mass ratio was 2 to 1, a ratio close to what was reported before,[20] it was 1 to 2 in one subject and it reached 4 to 1 in some cases. The variations in the fat mass to fat-free mass ratio changes in response to overfeeding were correlated with the changes in body weight and the coefficient reached 0.61 ($p < 0.01$).[16] In other words, about 37% of the variation in weight gain as a result of exposure to long-term overfeeding was associated with this dimension of nutrient partitioning. Those who gained more fat relative to fat-free tissues were the high gainers for body mass while those who gained relatively more lean tissues were the low gainers. Moreover, a significant genotype–overfeeding interaction effect ($p < 0.05$) was observed for the response of the fat mass (kg) to fat-free mass (kg) ratio (see Table 1).

If, on the average, 63% of the extra energy consumed was accounted for by the changes in fat mass and fat-free mass, one was left with 29,000 kcal as unaccountable. We measured the energy content of the feces for several days before and after the overfeeding treatment in 16 of the 24 subjects. There was no significant change in the amount of energy that was not absorbed during digestion. Thus, the remaining energy must in all likelihood be associated with the costs of protein (3,333 kcal) and fat (8,095 kcal) tissues gained and with increases in resting metabolic rate, thermic effect of food, standard postures, moving the body around and fidgeting. One must also entertain the possibility that increases in energy expenditure for postures, bodily activities, and fidgeting were present not only as a result of increases in body mass but perhaps also because of a shift in the daily pattern of activities toward a more energy-demanding profile even though subjects were kept under sedentary conditions. In fact, we were able to account for 91% of the 84,000 kcal energy surplus ingested by the 24 young adults of the study.[21] The remaining unexplained energy expenditure (about 9%) was likely accounted for by errors of measurement, errors in the assumptions made to estimate overall energy balance at the onset of the study, and by small differences in the daily pattern of activities over the duration of the overfeeding protocol.

The long-term overfeeding study also revealed that there was six times more variance for between-pairs than for within-pairs for the changes in upper body

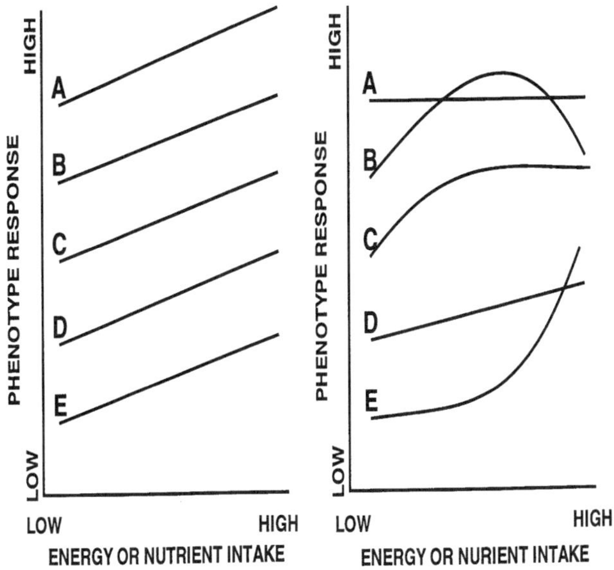

FIGURE 3. Changes in abdominal visceral fat in response to a 100-day overfeeding protocol in 12 pairs of young adult male identical twins. The within-pair resemblance in response is illustrated. (Reproduced from Ref. 16 by permission of the Massachusetts Medical Society.)

fat and abdominal visceral fat (determined by computerized tomography) when both changes were adjusted for the gain in total fat mass (Figure 3).[16] These observations indicate that some individuals are storing fat predominantly in selected fat depots primarily as a result of undetermined genetic characteristics. It also suggests that variations in regional fat distribution are more closely related to the genotype of the individuals than variations in body mass and overall body composition.

Metabolic Rates

Resting metabolic rate in absolute terms increased by about 10% with overfeeding ($p < 0.05$). However, the increase was only marginal ($p = 0.06$) when it was expressed by a unit of fat-free mass.[21,22] The intrapair resemblance for the changes in resting metabolic rate brought about by overfeeding was significant ($p < 0.05$) but it became nonsignificant when the changes in body mass or body composition were taken into account.

The thermic response to food, as assessed by indirect calorimetry for a period of 4 hr following the ingestion of a 1000-kcal meal of mixed composition, did not increase with overfeeding when resting metabolic rate was substracted from postprandial energy expenditure.[21] In contrast, postprandial energy expenditure and the total energy cost of weight maintenance increased significantly ($p < 0.05$), but the increments were mostly due to the gain in body mass.

Plasma, Insulin, and Glucagon

Fasting plasma glucose, insulin, and glucagon concentrations were significantly increased by overfeeding. During a 75-g oral glucose tolerance test, no major alteration in glucose tolerance was observed, and the insulin area under the curve was increased. During a meal test, insulin and glucagon areas under the curve were increased. The pre-overfeeding values of glucose, insulin, and glucagon (fasting and areas) were not correlated with the gains in body weight and fat mass. However, fasting glucagon before overfeeding was positively correlated with the gains in abdominal visceral fat and in femoral fat. The changes with overfeeding in the post-glucose insulin area were positively correlated with the changes in total subcutaneous fat, even after adjustment for total body fat gain.[23]

This study shows that a long-term overfeeding protocol with a mixed diet induced significant changes not only in glucose and insulin levels but also in glucagon concentrations. Fasting plasma glucose was slightly but significantly increased by overfeeding. This is in agreement with the results of the long-term overfeeding study by Sims et al.[2] and of other overfeeding studies of shorter duration.[24,25] No major change was observed in the glucose concentrations after a carbohydrate load or a mixed meal. In the long-term Vermont overfeeding study, increased glucose levels were noted after an oral glucose load during the period of weight gain but the values remained within the accepted normal range,[2] as in our data. Thus, it appears that increasing the duration of overfeeding, or the intensity of the nutritional stimulus for a short period of time, may lead to some decrement in glucose tolerance. However, this phenomenon is of moderate magnitude in healthy men, with no family history of diabetes or obesity.

Fasting insulin levels were increased, as well as the insulin areas during the oral glucose tolerance test and the meal test. Increased insulin concentrations in the fasting state is a general feature in overfeeding studies.[2,8,19,24,25] Our data showed an increase in the ratio of insulin area:glucose area which was observed after ingestion of glucose and after ingestion of a meal, suggesting that the prolonged nutritional challenge was associated with the development of some degree of insulin resistance. We also observed an increase in fasting plasma glucagon concentrations after the 100-day protocol. The insulin/glucagon ratio in the fasting state was not changed, indicating that insulin and glucagon were increased to the same extent. After ingestion of a mixed meal containing 15% protein, increased plasma glucagon concentrations were observed following the long-term caloric surplus. However, this effect was explained by the increase in fasting plasma glucagon levels.[23]

There were important differences between subjects for the changes in fasting insulin and glucagon. However, these individual differences were not distributed randomly among the 24 men, as shown by the high within pair similarity in the absolute response to overfeeding. There was about 6 times more variance between the pairs than within the pairs, with respect to fasting insulin concentration changes (Figure 4). Thus, some individuals were more

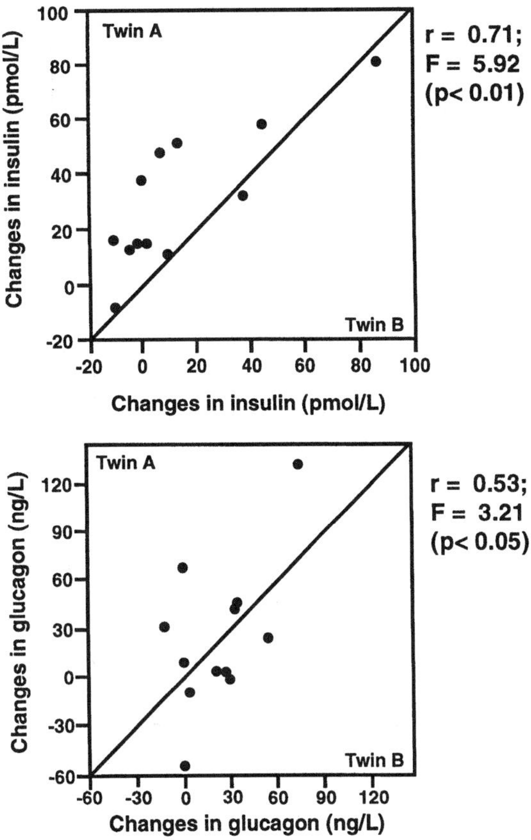

FIGURE 4. Changes in fasting plasma insulin and glucagon in response to a 100-day overfeeding protocol in 12 pairs of young adult male identical twins. The within-pair resemblance in response is illustrated. (Reproduced from Ref. 23 with permission.)

prone than others to exhibit an increase in fasting insulin levels. These results suggest that the genotype is an important determinant of the magnitude of the insulinemic response in times of positive energy balance. Fasting glucagon concentrations were also characterized by a significant within-pairs similarity in the response to overfeeding.[23]

Thyroid Hormones

In the Vermont overfeeding studies, Danforth et al.[26] showed that increased energy intake was associated with increased serum T_3 concentrations, whereas serum T_4 concentrations were unchanged. Increased mean T_3 concentrations with unchanged mean T_4 concentrations were also reported in several short-term overfeeding experiments[27-33] together with a decrease in r T_3 in two studies.[30-33] In spite of the frequently increased T_3 concentrations with short-term overfeeding, basal thyroid stimulating hormone (TSH) concentrations and the TSH response to thyroid releasing hormone (TRH) stimulation were found unchanged, suggesting a resetting of the hypothalamic-pituitary feedback system.[26,29]

In the long-term overfeeding study with MZ twins, basal plasma T_3 concentrations were increased at day 25 but not later. Basal plasma T_4 and free T_4 concentrations were unchanged. Basal plasma r T_3 concentrations were persistently decreased throughout the entire protocol, and the TSH response to TRH stimulation was persistently enhanced. The TSH response to TRH before overfeeding was negatively correlated with the changes in resting metabolic rate with overfeeding ($r = -0.52$, $p < 0.02$). No association was found between changes in basal plasma T_3 concentrations and changes in resting metabolic rate. However, changes in basal T_3 were positively related to changes in body weight ($r = 0.46$, $p < 0.05$).[34]

During overfeeding, there is some evidence of individual variation in the response of thyroid hormones to excess energy intake. In a short-term overfeeding study where individual results were reported, serum T_3 concentrations were increased by about 15 to 20% in some subjects while they remained unchanged in others.[25] Whether the individual differences in the response of thyroid hormones to excess energy intake could be genotype–dependent had been investigated earlier in one study.[10] No genotype-overfeeding interaction was found after a 3-week overfeeding protocol (1000 kcal/d) in 6 pairs of identical twins. Comparing the within-pair variance to the between-pair variance in response to long-term overfeeding revealed that there was about 2 to 2.5 more variance in between-pairs than within-pairs in the response of basal T_4 and free T_4 concentrations to overfeeding.[34] These results suggest that the genotype could be involved in basal T_4 and free T_4 variations during a prolonged positive energy balance period. Interestingly, higher heritability levels have been observed in a twin study for basal T_4 and free T_4 concentrations than for T_3 and TSH concentrations.[35]

Adipose Tissue Lipolysis

In the short-term overfeeding study with six pairs of MZ twins, we had observed a significant intrapair resemblance for the response of abdominal fat cell basal and catecholamine-stimulated lipolysis.[10] We repeated these measurements in the long-term overfeeding study again using adipose tissue samples obtained by a biopsy of the abdomen near the umbilicus.[36] Abdominal fat cell weight increased from 0.20 to 0.42 μg lipid/cell ($p < 0.0001$). No significant changes in basal and catecholamine-stimulated lipolysis were observed when lipolytic activity was expressed either per fat cell number or corrected for cell surface area. However, significant intrapair resemblance was found for the variation of basal and epinephrine-stimulated lipolysis in response to overfeeding. In both cases, there was about four to five times more variance for between-pairs than for within-pairs ($p \leq 0.05$) with intraclass correlation coefficients reaching about 0.6.[37]

LONG-TERM NEGATIVE ENERGY BALANCE STUDY

We have extended the investigation of the response to negative energy balance using the MZ study design over the long-term. Seven pairs of young adult male identical twins completed a negative energy balance protocol during which they exercised on cycle ergometers twice a day, 9 out of 10 days, for a period of 93 days while being kept on a constant daily energy and nutrient intake.[37] The total energy deficit caused by exercise above the estimated energy cost of body weight maintenance reached a mean of 58,095 kcal. Baseline energy intake was estimated over a period of 17 days preceding the negative energy balance period. Mean body weight loss was 5.0 kg ($p < 0.001$) and it was entirely accounted for by the loss of fat mass ($p < 0.001$). Fat-free mass was unchanged. Body energy losses reached 45,476 kcal ($p < 0.001$) which represented about 78% of the estimated energy deficit. Subcutaneous fat loss was slightly more pronounced on the trunk than on the limbs as estimated from skinfolds, circumferences, and computed tomography. The reduction in abdominal visceral fat was quite striking, from 81 cm^2 to 52 cm^2 ($p < 0.001$). At the same submaximal power output level, subjects oxidized more lipids than carbohydrates after the program as indicated by the changes in the respiratory exchange ratio ($p \leq 0.05$).[37]

Figure 5 depicts the changes in body weight and in abdominal visceral fat with the negative energy balance protocol. The intrapair resemblance in response is also illustrated. The F ratio of the between-pairs to the within-pairs variances reached about 6 for the body weight changes and 11 for the changes in visceral fat. Intrapair resemblance was observed for the changes in body weight ($p < 0.05$), fat mass ($p < 0.01$), percent fat ($p < 0.01$), body energy content ($p < 0.01$), sum of 10 skinfolds ($p < 0.01$), abdominal visceral fat ($p < 0.01$), fasting plasma triglycerides ($p < 0.05$), cholesterol ($p < 0.05$), maximal oxygen uptake ($p < 0.05$), and respiratory exchange ratio during submaximal work ($p < 0.01$).[37]

We concluded that even though there were large individual differences in response to the negative energy balance and exercise treatments, subjects with the same genotype were more alike in responses than subjects with different genotypes particularly for body fat, body energy, and abdominal visceral fat changes. High lipid oxidizers and low lipid oxidizers during submaximal exercise were also seen despite the fact that all subjects had experienced the same exercise and nutritional conditions for about three months.[37]

IMPLICATIONS OF THESE FINDINGS

The studies summarized in this review demonstrate that individual differences in response to chronic alterations in energy balance are ubiquitous. They are observed in a wide variety of phenotypes including body weight, body fat content, subcutaneous fat, visceral fat, postprandial energy expenditure, resting metabolic rate, adipose tissue metabolism, and insulin, glucagon, and thyroid

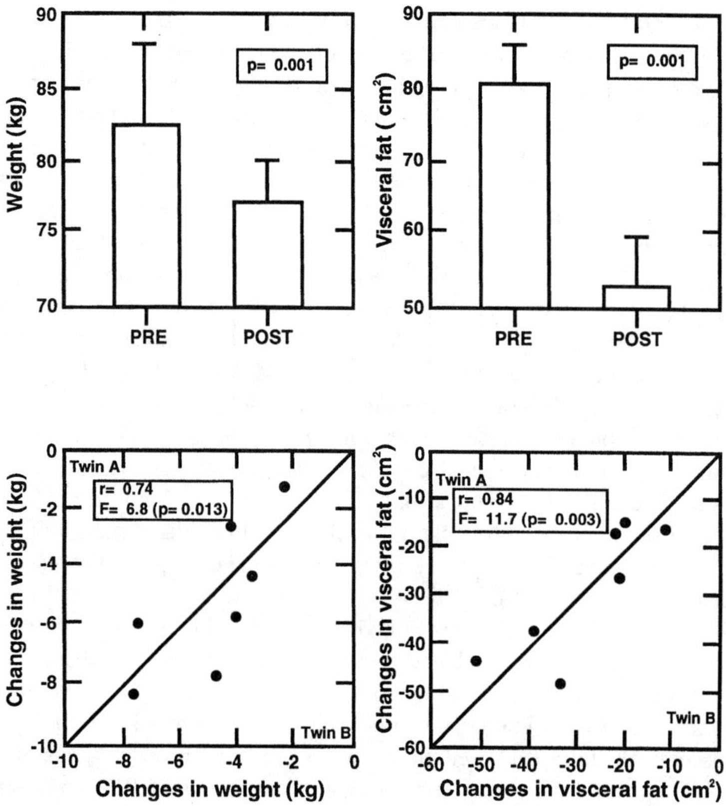

FIGURE 5. Changes in body weight (left panels) and abdominal visceral fat (right panels) in response to a 93-day negative energy balance and exercise protocol. Mean decrease and within-pair resemblance in response are illustrated. (Reproduced from Ref. 37 with permission.)

hormone levels. The comparison of the between-pairs and within-pairs heterogeneity reveal that members of the same twin pair (twin brothers) are generally more alike in their response to variation in energy balance than people who are not related by descent. These two series of observations support the notion that genotype–energy balance interaction effects are quite common.

As shown in Table 2, the mean MZ intrapair difference for body mass and body composition phenotypes attains about 25% of the mean change registered for the groups that took part in the long-term overfeeding or the long-term negative energy balance experiments, the within-pair heterogeneity being quite comparable for the two protocols. In general, the ratios of the variances of the between-pairs to the within-pairs ranged from about 2 to 10 depending on the phenotypes considered.

The genotype–energy balance interaction effects were quite consistent for the indicators of nutrient partitioning derived from the body composition

TABLE 2
Comparison of Mean Identical Twins Intrapair Difference in Response Between the Overfeeding and the Negative Energy Balance Studies

	Overfeeding Protocol[a]		Negative Energy Balance Protocol[b]	
	Mean Gain	Mean Intrapair	Mean Loss	Mean Intrapair
Body weight (kg)	8.1	1.8	5.0	1.4
Fat mass (kg)	5.4	1.6	4.9	1.1
Sum 10 skinfolds (mm)	53	15	52	14
Body energy (MJ)	222	63	191	41

a Based on 12 pairs of MZ twins.
b Based on 7 pairs of MZ twins.

changes. It would seem from the long-term overfeeding and negative energy balance studies summarized here that nutrient partitioning (proneness to gain or lose fat or fat-free tissues) is an important factor to explain the individuality in body mass changes with these experimental conditions. This is not a trivial factor as it acounted for at least 35 to 40% of the individual differences in body weight response in the two long-term protocols. This dimension of nutrient partitioning carries considerable implications for the risk of becoming obese or the protection from obesity as the kg of fat has an energy equivalent of 9300 kcal while the kg of lean tissue has only 1020 kcal. Chances are that a substantial proportion of those prone or resistant to obesity find themselves in this vulnerable or desirable position because of inherited or acquired differences in nutrient partitioning mechanisms.

Based on the results summarized here, we believe that it would now be useful to undertake similar intervention studies with a large number of singletons so that the full range of the heterogeneity in response to chronic alterations in energy balance can be documented. Such studies would also provide the opportunity, if the sample sizes were large enough, to investigate candidate genes and other molecular markers in an attempt to define the molecular basis of the genotype–energy balance interaction effects.

CONCLUSIONS

Some form of biological determinism can make one more susceptible to evolve toward a chronic positive or negative energy balance over prolonged periods of time. For a minority of individuals, say the high gainers in our experimental overfeeding studies or the low losers in the negative energy balance experiments, the genotype seems to contribute in major ways to the continuous battle these people have to wage in order to maintain body mass within acceptable limits. It is remarkable to see that such genotypes were observed in the intervention studies reported here despite the fact that the MZ twins pairs were not themselves at risk for obesity or obesity-related complications based on their own status and their family history.

The genes that are responsible for those individual differences in the sensitivity to alterations in energy balance remain to be identified. It would not be surprising if these genes turned out to be numerous considering the complexity of the biological systems that are involved in body weight homeostasis, nutrient partitioning, energy expenditure, and regulation of energy balance. A variety of genetic and molecular research strategies will be needed to begin to identify these genes and to delineate the nature and the extent of the genetic polymorphisms involved.

ACKNOWLEDGMENTS

The author would like to express his gratitude to all his colleagues who have contributed to these studies over the years, particularly: Drs. Angelo Tremblay, Germain Thériault, Eric Poehlman, Jean-Pierre Després, Denis Prud'homme, André Nadeau, Sital Moorjani, Paul Lupien, and Jean Dussault as well as Guy Fournier and Claude Leblanc. Thanks are also expressed to the personnel of the various laboratories that were involved in these studies. This research has been supported primarily by the National Institutes of Health (DK-34624) and the Natural Sciences and Engineering Research Council of Canada (G-0850).

REFERENCES

1. **Berdanier, C. D.,** Nutrient–gene interactions: today and tomorrow, *FASEB J.*, 8, 1, 1994.
2. **Sims, E. A. H., Goldman, R. F., Gluck, C. M., Horton, E. S., Kelleher, P. C., and Rowe, D. W.,** Experimental obesity in man, *Trans. Assoc. Am. Phys.*, 81, 153, 1968.
3. **Bouchard, C., Pérusse, L., and Leblanc, C.,** Using MZ twins in experimental research to test for the presence of a genotype–environment interaction effect, *Acta Genet. Med. Gemellol.*, 39, 85, 1990.
4. **Eaves, L. J.,** Human behavioural genetics, *Proc. R. Soc. Med.*, 69, 184, 1976.
5. **Plomin, R., Defries, J. C., and Loehlin, J. C.,** Genotype–environment interaction and correlation in the analysis of human behavior, *Psychol. Bull.*, 84, 309, 1977.
6. **Poehlman, E. T., Tremblay, A., Després, J. P., Fontaine, E., Pérusse, L., Thériault, G., and Bouchard, C.,** Genotype-controlled changes in body composition and fat morphology following overfeeding in twins, *Am. J. Clin. Nutr.*, 43, 723, 1986.
7. **Bouchard, C., Tremblay, A., Després, J. P., Poehlman, E. T., Thériault, G., Nadeau, A., Lupien, P., Moorjani, S., and Dussault, J.,** Sensitivity to overfeeding: the Québec experiment with identical twins, *Prog. Food Nutr. Sci.*, 12, 45, 1988.
8. **Poehlman, E. T., Tremblay, A., Fontaine, E., Després, J. P., Nadeau, A., Dussault, J., and Bouchard, C.,** Genotype dependency of the thermic effect of a meal and associated hormonal changes following short-term overfeeding, *Metabolism*, 35, 30, 1986.
9. **Tremblay, A., Poehlman, E. T., Nadeau, A., Dussault, J., and Bouchard, C.,** Heredity and overfeeding-induced changes in submaximal exercise VO_2, *J. Appl. Physiol.*, 62, 539, 1987.

10. **Poehlman, E. T., Després, J. P., Marcotte, M., Tremblay, A., Thériault, G., and Bouchard, C.,** Genotype dependency of adaptation in adipose tissue metabolism after short-term overfeeding, *Am. J. Physiol.,* 250, E480, 1986.
11. **Després, J. P., Poehlman, E. T., Tremblay, A., Lupien, P. J., Moorjani, S., Nadeau, A., Pérusse, L., and Bouchard, C.,** Genotype-influenced changes in serum HDL-cholesterol after short-term overfeeding in man: association with plasma insulin and triglyceride levels, *Metabolism,* 36, 363, 1987.
12. **Poehlman, E. T., Tremblay, A., Nadeau, A., Dussault, J., Thériault, G., and Bouchard, C.,** Heredity and changes in body composition and adipose tissue metabolism after short-term exercise training, *Eur. J. Appl. Physiol.,* 56, 398, 1987.
13. **Tremblay, A., Poehlman, E. T., Nadeau, A., Pérusse, L., and Bouchard, C.,** Is the response of plasma glucose and insulin to short-term exercise training genetically determined? *Horm. Metab. Res.,* 19, 65, 1987.
14. **Després, J. P., Moorjani, S., Tremblay, A., Poehlman, E. T., Lupien, P. J., Nadeau, A., and Bouchard, C.,** Heredity and changes in plasma lipids and lipoproteins after short-term exercise training in men, *Arteriosclerosis,* 8, 402, 1988.
15. **Poehlman, E. T., Tremblay, A., Nadeau, A., Dussault, J., Thériault, G., and Bouchard, C.,** Heredity and changes in hormones and metabolic rates with short-term training, *Am. J. Physiol.,* 250, E711, 1986.
16. **Bouchard, C., Tremblay, A., Després, J. P., Nadeau, A., Lupien, P. J., Thériault, G., Dussault, J., Moorjani, S., Pineault, S., and Fournier, G.,** The response to long-term overfeeding in identical twins, *N. Engl. J. Med.,* 322, 1477, 1990.
17. **Dériaz, O., Tremblay, A., and Bouchard, C.,** Non linear weight gain with long-term overfeeding in man, *Obes. Res.,* 1, 179, 1993.
18. **Norgan, N. G. and Durnin, J. V. G. A.,** The effect of 6 weeks of overfeeding on the body weight, body composition, and energy metabolism of young men, *Am. J. Clin. Nutr.,* 33, 978, 1980.
19. **Ravussin, E., Schutz, Y., Acheson, K. J., Dusmet, M., Bourquin, L., and Jéquier, E.,** Short-term, mixed-diet overfeeding in man: no evidence for "luxuskonsumption," *Am. J. Physiol.,* 249, E470, 1985.
20. **Forbes, G. B., Brown, M. R., Welle, S. L., and Lipinski, B. A.,** Deliberate overfeeding in women and men: energy cost and composition of the weight gain, *Br. J. Nutr.,* 56, 1, 1986.
21. **Tremblay, A., Després, J. P., Thériault, G., Fournier, G., and Bouchard, C.,** Overfeeding and energy expenditure in human, *Am. J. Clin. Nutr.,* 56, 857, 1992.
22. **Dériaz, O., Fournier, G., Tremblay, A., Després, J. P., and Bouchard, C.,** Lean-body mass composition and resting energy expenditure before and after long-term overfeeding, *Am. J. Clin. Nutr.,* 56, 840, 1992.
23. **Oppert, J. M., Nadeau, A., Tremblay, A., Després, J. P., Thériault, G., Dériaz, O., and Bouchard, C.,** Plasma glucose, insulin, and glucagon, before and after long-term overfeeding in identical twins, *Metabolism,* 44, 96, 1995.
24. **Olefsky, J., Crapo, P. A., Ginsberg, H., and Reaven, G. M.,** Metabolic effects of increased caloric intake in man, *Metabolism,* 24, 495, 1975.
25. **Welle, S. L., Seaton, T. B., and Campbell, R. G.,** Some metabolic effects of overeating in man, *Am. J. Clin. Nutr.,* 44, 718, 1986.
26. **Danforth, E., Horton, E. S., O'Connell, M., and Sims, E. H.,** Dietary induced alterations in thyroid hormone metabolism during overnutrition, *J. Clin. Invest.,* 64, 1336, 1979.
27. **Bray, G. A., Fisher, D. A., and Chopra, I. J.,** Relation of thyroid hormones to body weight, *Lancet,* 1, 1206, 1976.
28. **Davidson, M. B. and Chopra, I. J.,** Effect of carbohydrate and noncarbohydrate sources of calories on plasma 3,5,3′ triiodothyronine concentrations in man, *J. Clin. Endocrinol. Metab.,* 48, 577, 1979.
29. **Utiger, R. D.,** Differing thyrotropin responses to increased serum triiodothyronine concentrations produced by overfeeding and by triiodothyronine administration, *Metabolism,* 31, 180, 1983.

30. **O'Dea, K., Esler, M., Leonard, P., Stockigt, J. R., and Nestel, P.,** Noradrenaline turnover during under- and over-eating in normal weight subjects, *Metabolism,* 31, 896, 1984.
31. **Welle, S., O'Connell, M., Danforth, E., and Campbell, R.,** Decreased free fraction of thyroid hormones during carbohydrate overfeeding, *Metabolism,* 33, 837, 1984.
32. **Katzeff, H. L., O'Connell, M., Horton, E. S., Danforth, E. Jr., Young, J. B., and Landsbert, L.,** Metabolic studies in human obesity during overnutrition and undernutrition: thermogenic and hormonal response to norepinephrine, *Metabolism,* 35, 166, 1986.
33. **Burger, A. G., O'Connel, M., Scheidegger, K., Woo, R., and Danforth, E. Jr.,** Monodeiodination of triiodothyronine and reverse triiodothyronine during low and high calorie diet, *J. Clin. Endocrinol. Metab.,* 65, 829, 1987.
34. **Oppert, J. M., Dussault, J. H., Tremblay, A., Després, J. P., Thériault, G., and Bouchard, C.,** Thyroid hormones and thyrotropin variations during long-term overfeeding in identical twins, *J. Clin. Endocrinol. Metab.,* 79, 547, 1994.
35. **Meikle, A. W., Stringham, J. D., Woodward, M. G., and Nelson, J. G.,** Hereditary and environmental influences on the variation of thyroid hormones in normal male twins, *J. Clin. Endocrinol. Metab.,* 66, 588, 1988.
36. **Mauriège, P., Després, J. P., Marcotte, M., Tremblay, A., Nadeau, A., Moorjani, S., Lupien, P. J., Dussault, J., Fournier, G., Thériault, G., and Bouchard, C.,** Adipose tissue lipolysis after long-term overfeeding in identical twins, *Int. J. Obes.,* 16, 219, 1992.
37. **Bouchard, C., Tremblay, A., Després, J. P., Thériault, G., Nadeau, A., Lupien, P. J., Moorjani, S., Prud'homme, D., and Fournier, G.,** The response to exercise with constant energy intake in identical twins, *Obes. Res.,* 2, 400, 1994.

Chapter 6

NUTRIENT–GENE INTERACTIONS IN LIPOPROTEIN METABOLISM

Carolyn D. Berdanier

CONTENTS

INTRODUCTION

The composition and quantity of the food consumed by western societies has been linked to the development of hyperlipoproteinemia, atherosclerosis, and heart disease. Coronary vessel disease (CVD) is a complex disorder characterized by incremental losses in vascular elasticity due to the development of atherosclerosis. While high fat diets have been implicated in the atherosclerotic process, not all people (or animals) who consume such diets, die of CVD. There is a strong genetic linkage among those who develop CVD,[1] and this linkage is lacking in those who do not develop this disease. These observations of diet and genetic factors suggest, therefore, that CVD develops as a result of a nutrient–gene interaction. Ostensibly this seems reasonable, yet researchers seeking support for such an hypothesis have found a number of genes that are involved in the regulation of blood lipid levels and further, that the diet influence involves not only the amount of fat consumed but also the type of fat and the amount and type of carbohydrate. The nature of the diet effect on the expression of genes related to blood lipid ranges from influencing gene transcription to post-translational effects on the gene products. Linking

0-8493-9485-6/96/$0.00+$.50

TABLE 1
Ranges of Lipids in Humans

Lipid	Range (mg/dl)
Total cholesterol	146–270
Total triglycerides	54–321
VLDL cholesterol	3–56
LDL cholesterol	81–189
HDL cholesterol	29–62

Note: The range is defined as values which fall between the 5th and 95th percentile of values from healthy male subjects age 35–39 years.

these effects to the atherogenic process and subsequent disease has been a challenge to those researchers in this arena. While some progress has been made, we still do not have the details of this process in hand. A number of important biochemical markers have been identified and these will be described and discussed in subsequent sections of this chapter.

LIPEMIA

In the 1950s numerous epidemiological studies implicated the role of diet in heart disease. Among the industrialized "wealthy" nations of the world, heart disease is the leading cause of death.[2,3] Death from CVD has been correlated to blood lipid levels which were correlated to dietary fat intake. These correlations set the stage for a flurry of research activity directed towards identifying and describing blood lipids and relating diet intake to changes in these lipids. The lipids of interest were cholesterol and its esters and the triacylglycerides. Normal ranges of these lipids were established as guidelines for interpreting the blood assessments, shown in Table 1. Values in excess of these "normal" ranges were considered risk factors for CVD. That is, in people with higher than normal serum triglycerides and/or cholesterol, the risk of dying from CVD was increased. It was soon recognized that endogenous synthesis, transport, and storage were important determinants of blood lipid levels in addition to the dietary fat. This then led Fredrickson et al.[4] to develop a classification system for patients at risk for CVD due to their elevated blood lipids. This classification system is outlined in Table 2. Although outdated by our present knowledge about the function of the different lipid-carrying proteins and the various enzymes involved in fatty acid synthesis, fatty acid oxidation, triacylglyceride (triglyceride, TG) synthesis and hydrolysis and cholesterol synthesis, esterification and oxidation, this classification nonetheless laid the groundwork for viewing the various hyperlipidemias as genetic diseases. Ahrens and colleagues[5] as well as Fredrickson et al.[4] recognized that food lipid intake was probably not the main cause of elevated blood lipids and,

TABLE 2
An Early System for the Classification of Lipoproteimemias

Type	Key Features
I	Elevated chylomicrons
II	Elevated serum cholesterol
III	Elevated serum triglycerides
IV	Elevated serum cholesterol and triglycerides
V	Elevated serum cholesterol, triglycerides, and glucose intolerance

that in turn, elevated blood lipids did not always result in CVD. For example, genetic defects in chylomicron processing result in very high blood levels of both TG and cholesterol, yet this disorder is rarely associated with atherogenesis. Chylomicron remnant clearance defects, of which there are several, likewise are characterized by elevated blood lipids. These defects are more frequent in the general population (~ 1 in 5000) than are defects in chylomicron processing (1 in 1 million), yet people with these defects have very little increased risk for CVD.

In contrast, genetic defects in the fat-carrying proteins, the lipoproteins, and disorders in the cell surface receptors for these lipoproteins are associated with an increased risk for CVD. These defects are more prevalent in the general population than are the defects in chylomicron clearance and processing. The various defects are listed in Table 3 and the genes (so far identified) responsible for each of the processes involved in lipoprotein metabolism are listed in Table 4. Before mutations in these genes are discussed, a review of lipoprotein metabolism is in order.

TABLE 3
Mutations in Genes Relating to Lipoprotein Metabolism

Defective Gene	Frequency	Abnormality
Apo B (absent)	1:1,000,000	Abetalipoproteinemia
Apo B	1:1,000,000	Hypobetalipoproteinemia
LPL	1:1,000,000	Defective chylomicron clearance
Apo C-II	1:1,000,000	Defective chylomicron clearance
HTGL	?	Defective IDL clearance
Apo E	1:5000	Type III hyperlipoproteinemia
LDL receptor	1:500	Familial hypercholesterolemia
Apo B-100	1:500–1:1000	Familial defective apo B-100
LCAT, Type A	Rare	Familial lecithin:cholesterol transferase deficiency
LCAT, Type B	Rare	Fish eye disease (corneal opacity)
Apo A-I	1:1,000,000	Defective HDL production (Tangier's disease)
Apo A-I Milano	Rare	Reduced HDL cholesterol and apo A-I levels

TABLE 4
Genes Involved in Lipoprotein Metabolism

Gene	Chromosome Location	Function of Gene Product
Apo A-II	1	Transport protein in HDL
Apo B	2 p 23-24	Transport protein for chylomicrons
High density lipoprotein binding protein (HDLBP)	2 q 37	Binds HDL and functions in the removal of excess cellular cholesterol
Apo D	3	Transport protein similar to retinol-binding protein
Apo (a)	6	Abnormal transport protein for LDL
Hormone sensitive LPL		Lipase in adipocytes that acts on stored lipids and is activated by a catabolic hormone cascade
LPL	8 p 22	Lipase that acts on chylomicrons
Apo A-I	11	Transport protein for chylomicrons and HDL
Apo C-III	11	Transport protein for VLDL
Apo A-IV	11	Transport protein for chylomicrons
Hepatic LPL (HTGL)	15 q 21	Lipase that acts on IDL in liver. Reduces size of the HDL from HDL_2 to HDL_3
CETP	16 q 12-21	Participates in the transport of cholesterol from peripheral tissue to liver; reduces HDL size
LCAT	16 q 22.1	Participates in the reverse transport of cholesterol from peripheral tissues to the liver
LDL receptor	19	Proteins on the cell surface which bind LDL
Apo E	19	Protein remnant transferred from HDL to chylomicron remnant. Required for clearance of chylomicron remnant
Apo C-I	19	Transport protein for VLDL
Apo C-II	19	Chylomicron transport protein required cofactor for LPL activity

LIPOPROTEIN FORMATION AND PROCESSING

EXOGENOUS LIPID TRANSPORT

In the course of normal fat digestion and absorption, TG and cholesterol are processed in the intestinal epithelial cells into lipid-rich particles containing about 1% protein. These proteins are apo B, A-I, and apo IV. These particles, called chylomicrons, are present in the blood of feeding animals but usually are absent in starving animals. As the chylomicrons circulate, they acquire an additional protein, apo C-II. This additional protein is an essential cofactor for the recognition and hydrolysis of the chylomicron by the capillary endothelial enzyme, lipoprotein lipase (LPL). The LPL hydrolyzes most of the core triglycerides in the chylomicron, leaving a remnant that is rich in cholesterol and cholesterol esters. During the LPL catalyzed hydrolytic process, the excess surface compounds, i.e., the phospholipids and apolipoproteins B, A-I, and

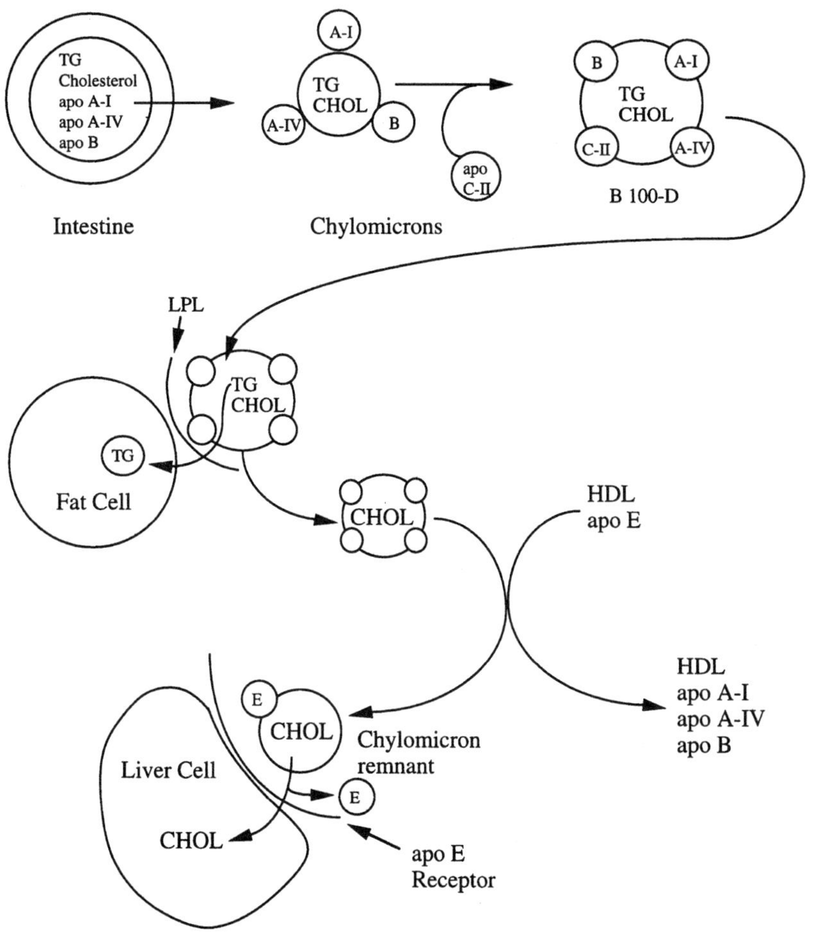

FIGURE 1. Schematic representation of the uptake, transport and disposal of dietary lipids. Abbreviations used: TG, triacyglycerides; CHOL, cholesterol and cholesterol esters; apo A-I, apo A-IV, apo B, apo E, polypeptide lipid carriers; HDL, high density lipoprotein.

A-IV are transferred to high density lipoproteins (HDL) and, in exchange, apo E is transferred from the HDL to the cholesterol ester-rich chylomicron remnant. These remnants are then cleared from the blood by the liver. On the hepatocyte is a lipoprotein receptor which recognizes apo E and this receptor plays an important role in remnant clearance. Figure 1 illustrates the process of exogenous lipid absorption, transport and clearance.

ENDOGENOUS LIPID TRANSPORT

Fatty acids, TG, cholesterol, cholesterol esters, and phospholipids are synthesized in the body and are transported from sites of synthesis to sites of use and storage. While the transport of these lipids is, in many instances, similar

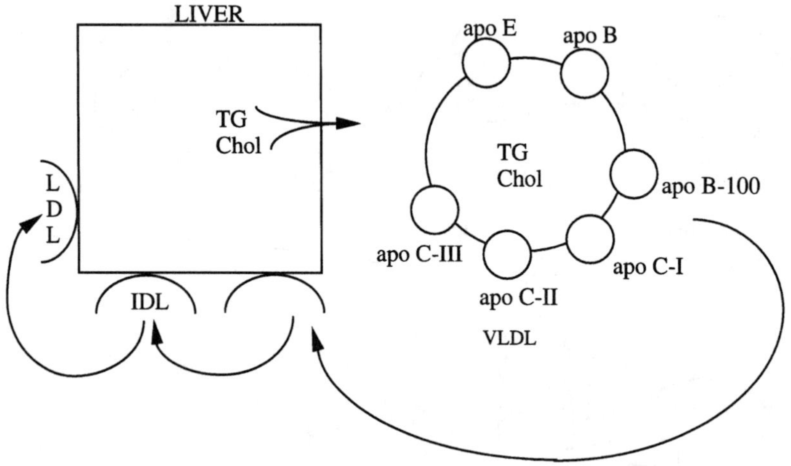

FIGURE 2. Lipid recycling–endogenous lipid transport and disposal. Very low density lipoproteins (VLDL) are secreted by the liver and circulated back to it where various receptors (shown as a half moon on the cell surface) bind and remove the lipid in a stepwise sequence. Abbreviations used: TG, triacylglyderides; CHOL, cholesterol and cholesterol esters; apo E, apo B, apo B-100, apo C-III, apo C-II, apo C-I, polypeptide lipid carriers; IDL, intermediate density lipoproteins; LDL, low density lipoproteins.

to that of the dietary lipids, there are differences in the processing and in some of the proteins involved. Endogenous fat transport involves the production and secretion of very low density lipoproteins (VLDL) by the liver. These lipid–protein complexes are rich in TG but also contain cholesterol. The polypeptides which transport these lipids comprise approximately 10% of the weight of the VLDL. They include the polypeptides apo B, B-100, apo C-I, apo C-II, apo C-III, and apo E. As mentioned, several of these polypeptides are also involved in exogenous lipid transport. Once the VLDLs are released by the hepatocyte, they are hydrolyzed by lipoprotein lipase and intermediate density lipoproteins (IDL) are formed. These are cleared from the circulation as they are recognized and bound to hepatic LDL receptors. The hepatic receptors recognize the apo E that is part of the IDL. Any of the IDL that escapes hydrolysis at this step is available for hydrolysis by the hepatic lipoprotein lipase. This hydrolysis leaves a cholesterol rich particle of low density (LDL). The LDL has apo B-100 as its polypeptide carrier and both hepatic and extrahepatic cells have receptors that recognize this polypeptide. Normally, about 70% of LDL is cleared by the LDL receptors and most of this is cleared by the liver. The endogenous fat transport and disposal system is diagramed in Figure 2. From the foregoing, it is apparent that considerable lipid recycling occurs in the liver. The VLDLs originate in the liver, which is the primary site for LDL disposal. However, other organs and tissues also participate in disposal, but their participation is minor compared to that of the liver.

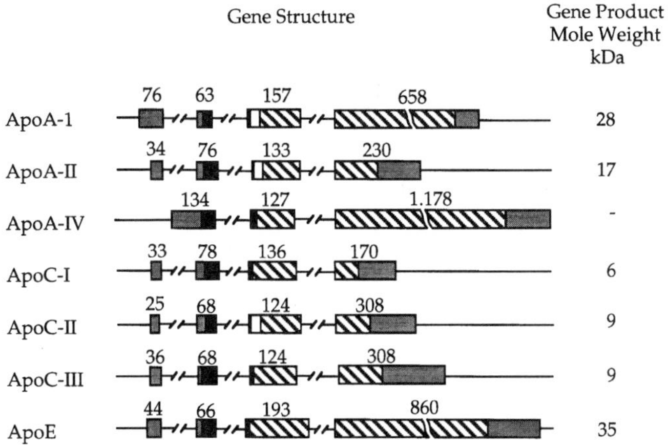

FIGURE 3. Gene structures and molecular weights of several of the lipid transport proteins. The 5,3′ untranslated region is shown as ▨, the leader sequence is shown as ■, the prosequence as □, the region coding for the mature protein is shown as ▧ . (From Rosseneu, M., and Lebeur, C., *FASEB J.*, 9, 768, 1995. With permission.)

GENETIC BASIS FOR LIPOPROTEINEMIAS

As mentioned, there are a number of proteins involved in the uptake, synthesis, and disposal of lipid. The transport proteins, apo A-I, apo A-II, apo A-IV, apo (a), apo B, apo D, apo C-I, apo C-II, apo C-III, apo D, and apo E have been identified (Table 4). In addition to these transport proteins, we have the proteins that are the important rate-limiting enzymes and receptors that are involved in lipoprotein processing: the lipoprotein lipase and the hepatic lipoprotein lipase, the lipoprotein receptors on the plasma membrane of the cells which receive and oxidize or store the transported lipids; and the rate-limiting enzymes of lipid synthesis and use. Included in this last category are the fatty acid synthase complex, acetyl CoA carboxylase, HMG CoA reductase, HMG CoA synthase, cholesterol ester transfer protein (CETP), fatty acid binding protein (FABP), lecithin–cholesterol acyltransferase (LCAT), cholesterol 7-hydroxylase, and the high density lipoprotein binding protein (HDLBP). The majority of these proteins have been isolated and studied in detail. With few exceptions, the genes which encode these products have been identified and mapped. Listed in Table 4 are the chromosomal locations of these genes and shown in Figure 3 are the gene structures of seven of the apolipoproteins. With this many genes involved in the uptake, synthesis, transport, and degradation of the circulating lipids, it is not surprising to find mutations in one or more of these genes which, when translated, result in an aberrant gene product which, in turn, expresses itself as a dysfunctional protein and consequently as one of the many diseases of lipid transport.[6] Not all of these diseases of lipid transport are associated with CVD. Table 3 lists these defects as well as estimates of their frequency. Defects in exogenous fat transport are manifested in several ways. Defective chylomicron formation due to the absence of the apo B gene leads to the absence of apo B and is characterized by fat malabsorption. This includes the malabsorption of fat soluble vitamins as well. Steatorrhea (fat-containing feces) as well as poor growth and the symptoms of fat soluble vitamin deficiency characterize the condition. Twenty different mutations have been identified in the gene for the apo B protein.[7] These mutations are inherited via an autosomal recessive mode and are characterized not only by fat malabsorption but also by acanthocytes, retinitis pigmentosa, and muscular neuropathies. To a large extent, these symptoms can be attributed to a relative deficiency of the fat soluble vitamins due to malabsorption. While a defect in the apo B gene can account for defective fat absorption, Wetterau[8] and Shoulders et al.[9] have suggested the presence of another mutation (in the microsomal triglyceride transfer protein) that also might result in fat malabsorption. This transfer protein is essential for apo B translocation and subsequent synthesis of chylomicrons. Defects in this transfer protein would impair apo B availability and chylomicron formation. A related condition is hypobetalipoproteinemia in which the apo B gene has mutated such that it is unable to carry the lipids from the gut into circulation. In this condition, very low levels of chylomicrons are found in the blood. Persons with this disorder are as rare (one in a million) as persons who lack the apo B gene. Fat malabsorption characterizes these people as well. The reason for the absence of the apo B has not been discovered.[10-14] Base deletion is the most likely explanation for the absence of the apo B gene product. The full base sequence need not be absent for the defect to occur; however, if only a few bases are deleted hypobetalipoproteinemia will be observed[12] rather than abetalipoproteinemia. In this circumstance the severity of the disease is related to the size of the truncated protein and whether it can associate with the lipids it must carry. The size of the truncated apo B can vary from apo B-9 (41 residues) to apo B-89 (4487 residues). Except in the case of the apo B-25, the result of a deletion of the entire exon 21, all the truncated forms reported to date are C-T transitions or base deletions.[7] These deletions can involve misaligned pairing deletion mechanisms.[15] Frameshift mutation can be compensated by a reading frame resaturation of the apo B gene.[16]

Familial hyperchylomicronemia is characterized by elevations in chylomicrons having both triglycerides and cholesterol. Hyperchylomicronemia was found to be due to mutations in the genes that encode the enzyme lipoprotein lipase (EC3.1.1.34) needed for the hydrolysis of chylomicrons.[17] This enzyme is a glycoprotein having an apparent monomeric molecular weight of about 60,000 Da on SDS gel electrophoresis and 48,300 Da by sedimentation-equilibrium ultracentrifugation.[18] The enzyme is linked to the endothelial cells of the capillary system.[17] Lipoprotein lipase is quite similar to hepatic triglyceride lipase, an enzyme found in the hepatic sinusoids. The main difference between the two lipases is that the lipoprotein lipase has a requirement for the lipid-carrying protein, apo C-II, for full activity whereas hepatic LPL does not. Mutations in the gene for apo C-II[19-21] can result in aberrant lipase activity because apo C-II serves as a cofactor in the lipoprotein lipase-catalyzed reaction.[19] Hepatic triglyceride lipase has no such requirement. Aberrations in the hepatic triglyceride lipase results in an accumulation of very low density lipoproteins rather than an accumulation of chylomicrons.[21] Mutations in the genes for LPL and apo C-II are very rare, occurring at a frequency of one in a million. The gene for LPL has been mapped to chromosome 8, while that for apo C-II has been mapped to chromosome 19, and the hepatic LPL to chromosome 15.[22] Other features of these disorders include an inflammation of the exocrine pancreas and eruptive xanthomas. Chylomicronemia does not appear to be atherogenic and CVD risk is not increased. The mutations in the LPL gene appear to be insertions or deletions[23] or due to aberrant splicing[24] while those in the apo C-II gene seem to be due to splice site mutations or small deletions.[25-27] Twenty-two mutations in the apo C-II gene have been reported.[7] With respect to the aberrant splicing of the LPL gene in three unrelated humans, Holzl et al.[24] reported a C→A mutation in position -3 of the acceptor splice site of intron 6 which caused aberrant splicing. The major transcript showed a deletion of exons 6 through 9 and amounted to about 3% of normal. Trace amounts of both a normally spliced LPL mRNA and a second aberrant transcript devoid of exon 7 were found. In one of these patients Holzl et al. found a 3′ splice mutation on one allele while on the other allele they found a missense mutation resulting in Gly 188→Glu substitution. All three subjects were classed as hyperchylomicronemic due to LPL deficiency.

The absence of LPL activity in certain tissues or in certain individuals can be attributed to mutations in lipoprotein lipase promoter region. Studies of tissue-specific expression of LPL in a variety of murine tissues by Gimble et al.[28] showed that cis-acting elements located within the –1824-bp of the 5′ flanking region was required for the expression of LPL. These include nuclear factors recognizing both the CCAAT box and the octamer sequence immediately flanking the transcriptional start site. Those tissues that have no LPL activity lack this promoter region. Since humans and mice have identical CCAAT and octamer sequences, one could suppose that humans having an intact LPL gene of normal sequence but lacking LPL activity might have a deficient or mutated promoter region. Although such has been suggested by Sparkes et al.,[22] proof that this might occur is presently lacking.

Mutations in the gene for hepatic lipoprotein lipase result in elevated blood levels of triglycerides and cholesterol, and these elevations are correlated to increased CVD risk. Hepatic LPL must be secreted by the hepatocyte into the sinusoids to function as a catalyst for the hydrolysis of core TG and surface phospholipids of chylomicron remnants, HDL and IDL. Through its activity, it augments the uptake of HDL cholesterol by the liver (reverse cholesterol transport) and is involved in the reduction of HDL size from HDL_2 to HDL_3. Hepatic LPL aids in the clearance of chylomicron remnants by exposing the apo E epitopes for enzymatic action. Missense mutations in the hepatic LPL gene include substitutions of serine for phenylalanine at amino acid position 267, threonine for methionine at position 383,[29] and asparagine for serine at position 291.[30] These mutations result in poorly secreted enzyme and thus the phenotypic expression of the mutation is low hepatic lipase activity. The frequency of the Asn 291 Ser mutation in a population having premature CVD has been reported as 5.2%.[30]

Defects in chylomicron remnant clearance are much more common than any of the above mutations.[31-38] Defective clearance due to mutations in the apo-E gene result in a lipemia known as Type III hyperlipoproteinemia. It is associated with premature atherosclerosis, and patients with these defects have high serum triglyceride levels as well as high serum cholesterol levels.[31] Xanthomas are found in nearly 75% of the population with these defects. The lipemia is responsive to energy restriction using diets which have 40% of energy from carbohydrate, 40% from fat, and 20% from protein. Weight loss is efficacious for most people with this defect. The apo E gene codes for the protein on the surface of the chylomicron remnant that is the ligand for receptor-mediated clearance of this particle. A number of mutations in the apo E gene have been reported and the phenotypes of these mutations are grouped into three general groups labeled E_2, E_3, and E_4. Those of the E_2 groups fail to bind the particles to the cell surface receptor for the chylomicron remnant. Those of the E_3 and E_4 groups have generally low remnant clearance rates. The apo E allele and phenotype frequency varies. The E_2 frequency is about 8%, the E_3 about 77%, and the E_4 about 15% of the total population with an apo E mutation. The incidence of apo E gene mutations is about 1% of the population. Since apo-E is involved in both endogenous and exogenous lipid transport and clearance, a faulty apo E gene is devastating. Mature human apo E is a 299 amino acid polypeptide. Apo E as well as other apolipoproteins contain 11 or 22 amino acid repeated sequences as one of their key features. These appear to encode largely amphipathic α helices which are needed for lipid binding.[35] There is a high degree of conservation among species of nucleotide sequences in the gene fragment that encodes the amino acid repeats.[35] The gene for apo E has been mapped to chromosome 19 as have the genes for apo C-I, apo C-II and LDLR. There appears to be a tight linkage among these genes which coordinates their expression.[36] Among the common mutations are amino acid substitutions at positions 112 and 158, while less frequent substitutions occur at other positions in the polypeptide chain. Several of these involve the exchange of neutral amino acids for acidic amino acids with the net result of alterations in polypeptide charge and subsequent inadequate binding to the appropriate cell surface receptors.[32] In a rare form of the disorder, the mutation is such that no useful apo E is formed.[33] Transgenic mice have been constructed with an apo E mutation that mimics apo E deficiency. These mice, like humans, develop hypercholesterolemia and increased susceptibility to atherosclerosis.[34] When these mice were fed low fat (5%) or high fat (16%) diets, a differential serum cholesterol pattern was observed: those fed the high fat diet had significantly higher levels of cholesterol, VLDL, and LDL than those fed a low fat or stock diet. The transgenic mice, even when fed the stock diet, had significantly higher levels of cholesterol, VLDL, and LDL than the normal control mice.

There was a gender difference as well. Male transgenic mice were less diet responsive in terms of their cholesterol levels than female transgenic mice. As mentioned, there is some sequence homology between humans and mice in the apo E DNA[35] and one could infer that these responses to dietary fat intake in mice could be observed in humans as well. While the above transgenic approach used the gene knockout paradigm (an extreme in the variants of apo E mutants), it nonetheless suggests that variation in the apo E genes could determine the responsiveness of humans with apo E defects to dietary manipulation. Indeed, Weintraub et al.[37] have reported that dietary fat clearance in normal subjects is regulated by genetic variance in apo E sequence. In addition, Jones et al.[38] have reported that the apo E genotype declares the response to cholesterol intake with respect to blood cholesterol levels and that this genotype influences cholesterol synthesis. Those subjects who respond poorly to an oral cholesterol challenge vis à vis blood cholesterol clearance had higher rates of cholesterol synthesis than those who could rapidly clear their blood of cholesterol after an oral challenge. One of the more interesting variants of the apo E is called the Milano variant. In this variant, blood lipid levels are elevated, but these elevations are not associated with an increased risk of atherosclerosis.

Cholesterol traffic is also controlled by the LDL receptor[39] and apo B-100.[40-42] Mutations in the gene for the LDL receptor or in the gene for apo B-100, the ligand for the LDL receptor, results in high serum levels of cholesterol. The former results in the disorder called familial hypercholesterolemia and occurs with a frequency of about 1 in 500. Familial hypercholesterolemia is associated with early death from atherosclerosis in man and related primates.[39-42] Dietary fat saturation affects transcription of LDL receptor mRNA in that feeding a diet containing saturated fat results in decreased LDL receptor mRNA compared to feeding an unsaturated fat diet. These results suggest that unsaturated fatty acids may interact with proteins that in turn serve as either cis- or trans-acting elements for this gene in much the same way as polyunsaturated fatty acids affect fatty acid synthetase gene expression.[43] Recently, Grossman et al. reported successful *ex vivo* gene therapy for a patient with an LDL receptor deficiency.[44] These researchers obtained a section of the patient's liver, perfused it with collagenase and then transfected the resultant hepatocytes with normal human LDL receptor using a retrovirus vector. The hepatocytes were then transfused back into the patient via the inferior mesenteric vein. This procedure was well tolerated and the patient showed improvement in cholesterol clearance. Her blood cholesterol levels fell and it was apparent that the "corrected" hepatocytes had grafted onto her remaining liver.

Mutations in the gene for apo B-100 result in a condition known as familial defective apo B-100 hypercholesterolemia.[45-47] In the general population this mutation occurs with a frequency of 1 in 500 as well. However, if only the individuals with high serum cholesterol levels are considered, the frequency is 1 in 50 to 1 in 100. Familial defective apo B-100 hypercholesterolemia is due to a mutation in the coding sequence of the apo B gene that changes the base sequence such that glutamine is substituted for arginine at bp 3500. This is in the LDL receptor binding region of the apo B protein and results in a binding affinity of less than 4% of normal.[45,47] Polymorphic variation in the genes for both the LDL receptor and the apo B-100 have been reported for mice[48] and this variation has provided the opportunity to identify the genetic and molecular constraints of lipoprotein gene expression. Both apo B and Apo E serve as ligands for the LDL receptor. In contrast to Apo E, apo B has little homology with the other apolipoproteins. Apo B in mice, is quite variable, and this variation imparts or confers a diet-responsive characteristic in inbred mouse strains vis à vis polypeptide sequence and activity.[48] That is, some mouse strains have reduced levels of plasma apo B when fed a high fat diet compared to controls fed a stock diet, while other mouse strains are unresponsive to diet vis à vis their plasma apo B-100 levels. Such polymorphism also exists in humans.[49] Apo B has been mapped to chromosome 2 while the LDL receptor has been mapped to chromosome 19. Apo B-48, a protein synthesized in the intestinal cell, is closely related to apo B-100. It differs in size (apo B-100 is a 550-kDa protein while apo B-48 is a 242-kDa protein) but has an identical amino-terminal amino acid. Both apo B-100 and apo B-48 are encoded by the same gene. In the intestine, the apo B primary transcript is co- or post-transcriptionally modified. This modification converts codon 2153 from a glutamine (CAA) to an in-frame, premature termination codon (UAA), thereby causing translation to terminate after amino acid 2152. This mRNA editing thus explains the difference in size of these two proteins.[50] If more of the apo B gene is deleted, hypocholesterolemia is observed.[51] This is because apo B is required for the transport of the chylomicrons from the intestine. If lacking, chylomicron formation is impaired and low serum cholesterol levels are observed. Both familial defective apo B-100 and familial hypercholesterolemia are characterized by high levels of LDL. Both are associated with CVD, but only the familial hypercholesterolemia is characterized by tendon xanthomas. Both are inherited as autosomal dominant disorders. Collectively, these mutations have a cumulative frequency of 1 in 250. However, because polymorphism in the apo B gene can and does occur, there is the possibility that collectively the frequency is much greater, that is, one in five. If this is the case, then the population variation in plasma cholesterol levels could be explained on the basis of these genetic differences alone, apart from those mutations that are associated with the rest of the genes which encode components of the lipid transport system.

Genetically determined abnormalities in LDL metabolism may also be due to mutation in the large glycoprotein apo (a). Apo (a) is a disulfide protein bonded to the apo B-100. It is thought to resemble plasminogen.[52] In fact, the genes which encode Lp(a) and plasminogen are very close to each other on the long arm of chromosome 6.[53] In general, LDLs containing apo (a) do not bind well to the LPL receptor, and people having significant amounts of apo (a) have a two- to threefold increase in CVD risk. Apo (a) is highly heritable and has been mapped to chromosome 6. Many individuals have little or no apo (a) and it has been suggested that those who have it are abnormal with respect to LPL activity.[53-56]

Several mutations in the genes which encode the endogenous lipid transport have been reported.[57-66] The reverse cholesterol transport pathway is part of this endogenous lipid transport system (see Figure 2). It involves the movement of cholesterol from peripheral tissues to the liver. The peripheral tissues cannot oxidize cholesterol and so must send it to the liver, whereupon it is prepared for excretion via cholesterol 7 α hydroxylase, as bile acids. This pathway uses the HDL to shuttle the cholesterol in this direction. HDL consists primarily of apo A-1 and cholesterol, which is esterified by the enzyme lecithin–cholesterol acyltransferase. Mutations in the LCAT gene have been reported[57] and results in one of two diseases: familial LCAT deficiency and fish eye disease (FED). Thirteen different mutations have been identified and in one, a single T→A transversion in codon 252 in exon 6, converting met (ATG) to Lys (AAG), was observed.[57] Three unrelated families were found to have this mutation; however, the severity of their disease varied. In these families, no other mutation in LCAT was observed. Of the remaining 12 LCAT mutations, 10 had point mutations, 3 had frameshifts, and 1 had a three-base insertion which maintained its reading frame. For FED, three mutations are known. This disorder is less serious than familial LCAT deficiency and is characterized by dyslipoproteinemia and corneal opacity. LCAT activity is 15% of normal and there is a reduced level of HDL in the plasma. In contrast, familial LCAT deficiency is characterized by a variety of symptoms including lipoprotein abnormalities, renal failure, premature atherosclerosis, reduced levels of plasma cholesterol esters, and high plasma levels of cholesterol and lecithin. The LCAT enzyme requires apo A-I as a cofactor. If there is a mutation in the gene for apo A-I, defective HDL production results. This is a rare mutation and its frequency is estimated as one in a million. Individuals with this defect have premature CVD, corneal clouding, and very low HDL levels. Plasma apo A-I has a variety of charge isoforms with similar antigenicity, amino acid composition, and peptide maps, yet all of these forms originate from the same gene.[58] Humans, baboons, African green monkeys, and cynomolgus monkeys have been studied and there are species differences in hepatic and intestinal apo A-I production.[59-62] In all instances, differences in apo A-I were reported between the intestine and the liver. In the liver, there was a twofold higher level of apo A-I mRNA than in the intestine[59] and the abundance of this mRNA was species-specific. The apo A-I gene is regulated at the level of transcription and a portion of the species-specific difference in apo A-I gene expression is due to a sequence divergence in the 5′ regulatory region including the exon/intron 1 of the apo A-I gene.[60] The capacity to produce HDLs is both genetically controlled and tissue-specific, and probably explains why some genotypes respond normally to a high fat diet by producing more HDL while other genotypes become hypercholesterolemic under these same conditions.[61] Attempts to create a transgenic mouse expressing a human apo A-I gene have not been fully successful but have provided additional information about the relationship of apo A-I to HDL size.[62] The human apo A-I gene was inserted into the mouse and in these mice, both the mouse and the human genes were expressed. This dual expression suggests a species difference in the control of this expression. In other words, the control points differed and this resulted in a broader spectrum of HDL particles. Of interest is the observation[63] that when these transgenics were made, the naturally occurring atherosclerosis characteristic of the host strain[64] was ameliorated. This could well be attributed to the increased production of HDL and thence to increased cholesterol disposal. The host strain was C57BL/6 and atherogenesis was induced by feeding a high fat diet.[64] Other mouse strains do not become atherogenic when fed this high fat diet and thus this mouse strain is said to have an atherosclerosis gene (Ath-1).

Defective HDL metabolism due to a mutation in the apo A-I gene results in a rare autosomal dominant disorder described in a small group of villagers in Italy. Affected individuals have reduced levels of HDL cholesterol and apo A-I levels but have no increased risk of CVD. The disorder, named Apo A-I Milano is due to a point mutation in the apo A-I gene, changing codon 173 so that cysteine is used instead of arginine. Normal apo A-I has no cysteine so this change has effects (because of the sulfide group in the cysteine) on the apo A-I structure.

Defective lipoprotein processing has already been discussed with respect to lipoprotein lypase (LPL), long chain acyl transferase (LCA), and apo C-II deficiencies. A deficient cholesterol ester transport protein has been reported due to a mutation of the gene for this protein located on chromosome 16. A mutation in this gene has been used to explain the atherogenicity of high fat diets in primates,[65,66] but to date no evidence of such a mutation in humans has been put forward.

DIET EFFECTS ON GENE EXPRESSION

In the preceding section, mention was made of how the level or saturation of dietary fat could have effects on specific mRNAs or on the levels of certain of the transport proteins. The question that arises is how dietary fat can have this effect. Is it strictly substrate induction? The answer is probably "no" because not all genotypes respond in the same way to the same extent. What then, is the role of dietary ingredients in the expression of these genes that encode the various components of the lipid transport system? Editing of the apo B-100 and apo B-48 genes is enhanced by dietary carbohydrate. A number of genes have carbohydrate response elements and it is possible that these genes also have them. A carbohydrate response element has been identified in the apo E gene.[67] Carbohydrate influences mRNA stability and RNA processing.[68] Similarly, the gene which encodes the seven-enzyme complex of mammalian fatty acid synthetase has a lipid response unit in its structure as well.[43,68,69] No doubt we will learn that the lipid transport genes will be found to have a lipid response element as well.

Dreon et al.[70] studied male human subjects having either large LDL (pattern A) or smaller LDL (pattern B) particles. Humans with pattern B LDL particles have a higher level of fasting triglycerides and lower levels of HDL particles than humans with pattern A LDL particles. The former group also has a higher risk of CVD. At any rate, Dreon et al. evaluated the LPL response of these subjects to a low fat diet (24% of energy from fat) and compared these responses to those of subjects with pattern A (normal subjects) and to the LDL patterns of these subjects when consuming a higher fat diet (46% of energy from fat). A total of 105 men were studied. They found that those subjects having the pattern B phenotype had a significant reduction in serum cholesterol and apo B levels when consuming the low fat diet. Pattern A subjects had no such cholesterol response and further, these subjects had a small rise (from 86 ± 4 to 123 ± 7 mg/dl) in serum triglycerides. The result of this study illustrates that the effectiveness of modifying the fat intake of humans in an effort to modify serum lipid levels depends upon the genotype of the human. For the majority of the subjects in the Dreon study, the diet effect was nonexistent, but for 18 of the 105 men, the low fat diet resulted in a significant reduction in apo B levels and serum cholesterol. A reduction in apo B means that fewer chylomicrons were being formed. The low fat diet provided less fat for absorption and thus likely had effects on apo B gene translation. This suggests that the apo B gene has a lipid response element. In the pattern B subclass of subjects there may be overexpression of the lipid-carrying proteins of gut origin that could be attributed to the amount of fat consumed. This would occur if each of the genes had a lipid response element that served to promote the expression of these genes in individuals consuming a high fat diet. Those individuals not having the pattern B response possibly lack this lipid response element. Other instances of a gene having a lipid response element have been reported. Over-transcription of the fatty acid synthase gene (which has a lipid response element) has been shown to occur in genetically obese Zucker rats.[71] Mouse apo A-IV gene expression in C57BL/6 mice, mice shown by Paigen et al.[64] to develop atherosclerosis, is induced by high fat feeding. In normal mice this does not occur.[72] The apo A-IV gene consists of three exons and two introns. The introns separate the evolutionarily conserved and functional polypeptide domains. Intron 1 divides most of the apo A-IV signal peptide from the amino terminus of the mature plasma protein. The second intron separates a highly conserved variant amphipathic peptide repeat from the remainder of the mature apo A-IV protein. The 5′ flanking region has several features: variant TATA and CAT box sequences (TTTAAA and CCAACG); five G-rich direct repeats of ten nucleotides and a short inverted repeat. The variants found in this region may determine whether dietary fat can induce apo A-IV gene expression. Some may constitute a lipid response element while other sequence variants may not be lipid responsive. Further studies[73] showed that the genetic control of apo A-IV mRNA levels involves both cis-acting elements and genetically distinct trans-acting factors. The cis-acting elements (nucleic acid sequences) and the trans-acting factors (specific proteins) determine both the rate of mRNA transcription and its turnover. Transcription requires the recognition of promoter sequences by protein transcription factors. These factors may bind lipid and be part of the lipid response unit or the lipid response unit may be entirely separate. This has not been determined but merely speculated.

Dietary lipids even in the absence of direct effects on transcription and translation do influence the phenotypic expression of specific genotypes either because of overconsumption or because they have effects on certain of the hormones, i.e., insulin or the steroid hormones or the catabolic hormones that regulate or influence lipid synthesis, oxidation, and storage. For example, Chen et al.[74] showed that in diabetic mice, insulin regulated the transcription of the apo C-III gene and that this regulation in turn affected serum triglyceride levels. Triglyceride levels fell in the insulin-treated diabetic mice subsequent to an insulin-induced increase in apo C-III mRNA level. In turn, Chen et al. were able to show that this mRNA increase was due to a specific effect of insulin on apo C-III gene transcription.

SUMMARY

A large number of mutations in the various genes which encode the polypeptides responsible for the transport of lipids in the blood have been reported. In some instances the mutations result in lipid malabsorption and low blood lipid levels. However, the majority of the mutations have the reverse effect. Elevated serum lipids, either triglycerides or cholesterol or both, are characteristic of the disorder. Within the latter group of mutations are those whose expression can be manipulated by the type and amount of fat in the diet. If these mutations can be identified in humans having elevated blood lipids, reduction in dietary fat can be efficacious in lowering blood lipids and perhaps retard the development of atherosclerosis and CVD. Not all mutations will respond in this way. The success of such treatment depends largely on the correct identification of genotype and whether the gene in question has a lipid (or perhaps carbohydrate) response element in its promoter sequence. Given the variation in response of humans to dietary fat manipulation, it is clear that one dietary recommendation vis à vis fat intake is inappropriate for all genotypes.[75,76]

REFERENCES

1. **Barrett-Conner, E., and Khaw, K.-T.,** Family history of heart attack as an independent predictor of death due to cardiovascular disease, *Circulation,* 69, 1065, 1984.
2. **Brunzell, J. D. and Austin, M. A.,** Individuality, hyperlipidemia, and premature coronary artery disease, in *Genetic Variation and Nutrition,* Simopoulos, A. P. and Childs, B., Eds., S. Karger, Basel, 1990, 72.

3. **Heinemann, L., Bulyzhenkov, V., and Berg, K.**, Environmental and genetic risk factors for coronary heart disease, in *Genetic Variation and Nutrition*, Simopoulos, A. P. and Childs, B., Eds., S. Karger, Basel, 1990, 84.
4. **Fredrickson, D. S., Levy, R. I., and Lees, R. S.**, Fat transport in lipoproteins — an integrated approach to mechanisms and disorders, *N. Engl. J. Med.*, 276, 273, 1967.
5. **Ahrens, E. H. Jr., Hirsch, J., Oette, K., Farquhar, J. W., and Stein, Y.**, Carbohydrate induced and fat induced lipemia, *Trans. Assoc. Am. Physicians*, 74, 134, 1961.
6. **Breslow, J. L.**, Genetic basis of lipoprotein disorders, *J. Clin. Invest.*, 84, 373, 1989.
7. **Rosseneu, M. and Labeur, C.**, Physiological significance of apolipoprotein mutants, *FASEB J.*, 9, 768, 1995.
8. **Wetterau, J. R., Aggerbeck, L. P., Bouma, M. E., Eisenberg, C., Munck, A., Hermier, M., Schmitz, J., Gay, G., Rader, D. J., and Gregg, R. E.**, Absence of microsomal triglyceride transfer protein in individuals with abetalipoproteinemia, *Science*, 258, 999, 1992.
9. **Shoulders, C. C., Brett, D. J., Bayliss, J. D., Narcisi, T. M. E., Jarmuz, A., Grantham, T. T., Leoni, P. R. D., Bhattacharya, S., Pease, R. J., Cullen, P. M., Levi, S., Byfield, P. G. H., Purkis, P., and Scott, J.**, Abetalipoproteinemia is caused by defects in the gene encoding the 97 kDa subunit of a microsomal triglyceride transfer protein, *Hum. Mol. Genet.*, 2, 2109, 1993.
10. **Herbert, P. N., Assmann, G., Gotto, A. M., and Fredrickson, D. S.**, Familial lipoprotein deficiency: abetalipoproteinemia, hypobetalipoproteinemia and Tangier disease, in *The Metabolic Basis of Inherited Disease*, 5th ed., Stanbury, J. B., Wyngaarden, J. B., Fredrickson, D. S., Goldstein, J. L., Brown, M. D., Eds., McGraw-Hill, New York, 1983, 589.
11. **Leppert, M., Breslow, J. L., Wu, L., Hasstedt, S., O'Connell, P., Lathrop, M., Williams, R. R., White, R., and Lalouel, J. M.**, Inference of a molecular defect of apolipoprotein B in hypobetalipoproteinemia by linkage analysis in a large kindred, *J. Clin. Invest.*, 82, 847, 1988.
12. **Huang, L-S., Ripps, M. E., Korman, S. H., Deckelbaum, R., and Breslow, J. L.**, Apo B gene exon 21 deletion in familial hypobetalipoproteinemia (HBLP), *J. Biol. Chem.*, 264, 11394, 1989.
13. **Talmud, P. J., Lloyd, J. K., Muller, D. P. R., Collins, D. R., Scott, J., and Humphries, S.**, Genetic evidence from two families that the apolipoprotein B gene is not involved in abetalipoproteinemia, *J. Clin. Invest.*, 82, 1803, 1988.
14. **Huang, L. S., Janne, P., deGraff, J., Deckelbaum, R., Kayden, H., and Breslow, J. L.**, Exclusion of the absolute linkage between apo B gene and abetalipoprotein-emia, *Am. J. Hum. Genet.*, 41, 622, 1989.
15. **Groenwegen, W. A., Krul, E. S., and Schonfeld, G.**, Apolipoprotein B-52 mutation associated with hypobetalipoproteinemia is compatible with a misaligned pairing deletion mechanism, *J. Lipid Res.*, 34, 971, 1993.
16. **Linton, M. F., Pierotti, V., and Young, S. G.**, Reading frame restoration with an apolipoprotein B gene frameshift mutation, *Proc. Natl. Acad. Sci. U.S.A.*, 89, 11431, 1992.
17. **Greten, H. and Beil, F. U.**, Lipase deficiencies, *J. Inher. Metab. Dis.*, 11, (Suppl. 1), 29, 1988.
18. **Augustin, J. and Greten, H.**, Hepatic triglyceride lipase and plasma lipids, *Prog. Biochem. Pharmacol.*, 15, 5, 1979.
19. **LaRosa, J. C., Levy, R. I., Herbert, P., Lux, S. E., and Fredrickson, D. S.**, A specific apolipoprotein activator for lipoprotein lipase, *B.B.R.C.*, 41, 57, 1970.
20. **Breckenridge, W. C.**, Hypertriglyceridemia associated with deficiency of apolipoprotein C-II, *N. Engl. J. Med.*, 298, 1265, 1978.
21. **Breckenridge, W. C., Little, J. A., Alaupovic, P., Wang, C. S., Kuksis, A., Kakis, G., Lindgrene, F., and Gardner, G.**, Lipoprotein abnormalities associated with a familial deficiency of hepatic lipase, *Atherosclerosis*, 45, 161, 1982.

22. **Sparkes, R. S., Zollman, S., Klisak, I., Kirchgessner, T. G., Komaromy, M. C., Mohandas, T., Schotz, M. C., and Lusis, A. J.**, Human genes involved in lipolysis of plasma lipoproteins: mapping of loci for lipoprotein lipase to 8p22 and hepatic lipase to 15q21, *Genomics*, 1, 138, 1987.
23. **Langlois, S., Deeb, S., Brunzell, J. D., Kastelein, J. J., and Hayden, M. R.**, A major insertion accounts for a significant proportion of mutations underlying human lipoprotein lipase deficiency, *Proc. Soc. Nat. Acad. Sci. U.S.A.*, 86, 948, 1989.
24. **Holzl, B., Haber, R., Paulweber, B., Patsch, J. R., and Sandhofer, F.**, Lipoprotein lipase deficiency due to a 3′ splice site mutation in intron 6 of lipoprotein lipase gene, *J. Lipid Res.*, 35, 2161, 1994.
25. **Fojo, S. S., Beisiegel, U., Beil, U., Higuchi, K., Bojanovski, M., Gregg, R. E., Greten, H., and Brewer, H. B.**, Donor splice site mutation in the apolipoprotein (apo) C-II gene (apo C-II Hamburg) of a patient with apo C-II deficiency, *J. Clin. Invest.*, 82, 1489, 1988.
26. **Cox, D. W., Wills, D. E., Quan, F., and Ray, P. A.**, A deletion of one nucleotide results in functional deficiency of apolipoprotein C-II (apo C-II, Toronto), *J. Med. Genet.*, 25, 649, 1988.
27. **Fojo, S. S., Stalenhoef, A. F. H., Marr, K., Greg, R. E., Ross, R. S., and Brewer, H. B.**, A deletion mutation in the apo C-II gene (apo C-II, Nijmegen) of a patient with a deficiency of apolipoprotein C-II, *J. Biol. Chem.*, 263, 17913, 1988.
28. **Gimble, J. M., Hua, X., Wanker, F., Morgan, C., Robinson, C., Hill, M. R., and Nadon, N.**, In vivo and in vitro analysis of murine lipoprotein lipase gene promotor: tissue specific expression, *Am. J. Physiol.*, 268, E213, 1995.
29. **Durstenfeld, A., Ben-Zeev, O., Reue, K., Stahnke, G., and Doolittle, M. H.**, Molecular characterization of human lipase deficiency. In vitro expression of two naturally occurring mutations, *Arterioscler. Thromb.*, 14, 381, 1994.
30. **Reymer, P. W. A., Gagne, E., Groenemeyer, B. E., Zhang, H., Forsyth, I., Jansen, H., Seidell, J. C., Kromhout, D., Lie, K. E., Kastelein, J., and Hayden, M. R.**, A lipoprotein lipase mutation (Asn 291 Ser) is associated with reduced HDL cholesterol levels in premature atherosclerosis, *Nat. Genet.*, 10, 28, 1995.
31. **Morganroth, J., Levy, R. I., and Fredrickson, D. S.**, The biochemical, clinical, and genetic features of type III hyperlipoproteinemia, *Ann. Int. Med.*, 82, 158, 1975.
32. **Hui, D. Y., Innerarity, T. L., and Mahley, R. W.**, Defective hepatic lipoprotein receptor binding of β-very low density lipoproteins from Type III hyperlipoproteinemic patients. Importance of apoprotein E, *J. Biol. Chem.*, 259, 860, 1984.
33. **Ghiselli, G., Schaefer, E. J., Gascon, P., and Brewer, H. B.**, Type III hyperlipoproteinemia associated with apolipoprotein E deficiency, *Science*, 214, 1239, 1981.
34. **van Ree, J. H., van den Broek, W. J. A. A., Dahlmans, V. E. H., Groot, P. H. E., Vidgeon-Hart, M., Frants, R. R., Wieringa, B., Havekes, L. M., and Hofker, M. H.**, Diet induced hypercholesterolemia and atherosclerosis in heterozygous apolipoprotein E-deficient mice, *Atherosclerosis*, 111, 25, 1994.
35. **Rajavashisti, T. B., Kaptein, J. S., Reue, K. L., and Lusis, A. J.**, Evolution of apolipoprotein E: mouse sequence and evidence for an 11-nucleotide ancestral unit, *Proc. Natl. Acad. Sci. U.S.A.*, 82, 8085, 1985.
36. **Lusis, A. J., Heinzmann, C., Sparkes, R. S., Scott, J., Knott, T. J., Geller, R., Sparkes, M. C., and Mohandas, T.**, Regional mapping of human chromosome 19: organization of genes for plasma lipid transport (APO CI, -C2 and -E and LDLR) and of the genes C3, PEPD, and GP1, *Genetics*, 83, 3929, 1986.
37. **Weintraub, M. S., Eisenberg, S., and Breslow, J. L.**, Dietary fat clearance in normal subjects is regulated by genetic variation in apolipoprotein E, *J. Clin. Invest.*, 80, 1571, 1987.
38. **Jones, P. J. H., Main, B. F., and Frolich, J. J.**, Response of cholesterol synthesis to cholesterol feeding in men with different apolipoprotein E genotypes, *Metabolism*, 42, 1065, 1993.

39. **Brown, M. S. and Goldstein, J. L.,** Lipoprotein receptors in the liver control signals for plasma cholesterol traffic, *J. Clin. Invest.,* 72, 743, 1983.
40. **Kusumi, Y., Scanu, A. M., McGill, H. C., and Wissler, R. W.,** Atherosclerosis in a rhesus monkey with genetic hypercholesterolemia and elevated plasma Lp(a), *Atherosclerosis,* 99, 165, 1993.
41. **Scanu, A. M., Khalil, A., Neven, L., Tidore, M., Dawson, G., Pfaffinger, D., Jackson, E., Carey, K. D., McGill, H. C., and Fless, G. M.,** Genetically determined hypercholesterolemia in a rhesus monkey family due to a deficiency of the LDL receptor, *J. Lipid Res.,* 29, 1671, 1988.
42. **Fox, J. C., McGill, H. C., Carey, K. D., and Getz, G. S.,** In vivo regulation of hepatic LDL receptor mRNA in the baboon. Differential effects of saturated and unsaturated fat, *J. Biol. Chem.,* 262, 7014, 1987.
43. **Clarke, S. D. and Jump, D. B.,** Regulation of hepatic gene expression by dietary fats: a unique role for polyunsaturated fatty acids, in *Nutrition and Gene Expression,* Berdanier, C. D. and Hargrove, J. L., Eds., CRC Press, Boca Raton, FL, 1993, 227.
44. **Grossman, M., Raper, S. E., Kozarsky, K., Stein, E. A., Engelhardt, J. F., Muller, D., Lupien, P. J., and Wilson, J. M.,** Successful ex vivo gene therapy directed to liver in a patient with familial hypercholesterolemia, *Nat. Genet.,* 6, 335, 1994.
45. **Innerarity, T. L., Balestra, M. E., Arnold, K. S., Mahley, R. W., Vega, G. L., Grundy, S. M., and Young, S. F.,** Isolation of defective receptor-binding low density lipoproteins from subjects with familial defective apolipoprotein B-100, *Atherosclerosis,* 8, 551a, 1988.
46. **Soria, L. F., Ludwig, E. H., Clark, H. R. G., Vega, G. L., Grundy, S. M., and McCarthy, B. J.,** Association between a specific apolipoprotein B mutation and familial defective apolipoprotein B-100, *Proc. Natl. Acad. Sci. U.S.A.,* 6, 587, 1989.
47. **Innerarity, T. L., Mahley, R. W., Weisgraber, K. H., Bersot, T. P., Krauss, R. M., Vega, G. L., Grundy, S. M., Friedl, W., Davignon, J., and McCarthy, B. J.,** Familial defective apolipoprotein B-100: a mutation of apolipoprotein B that causes hypercholesterolemia, *J. Lipid Res.,* 31, 1337, 1990.
48. **Lusis, A. J., Taylor, B. A., Quon, D., Zollman, S., and LeBoeuf, R. C.,** Genetic factors controlling structure and expression of apolipoproteins B and E in mice, *J. Biol. Chem.,* 262, 7594, 1987.
49. **Mehrabian, M., Sparkes, R. S., Mohandas, T., Klisak, I. J., Schumaker, V. N., Heinzman, C., Zollman, S., Ma, Y., and Lusis, A. J.,** Human apolipoprotein B: chromosomal mapping and DNA polymorphisms of hepatic and intestinal species, *Somatic Cell Mol. Genet.,* 12, 245, 1986.
50. **Bostrom, K., Garcia, Z., Poksay, K. S., Johnson, D. F., Lusis, A. J., and Innerarity, T. L.,** Apolipoprotein B mRNA editing, *J. Biol. Chem.,* 265, 22446, 1990.
51. **Young, S. G., Northey, S. T., and McCarthy, B. J.,** Low plasma cholesterol levels caused by a short deletion in the apolipoprotein B gene, *Science,* 241, 591, 1988.
52. **McLean, J. W., Tomlinson, J. E., Kuang, W-J., Eaton, D. L., Chen, E. Y., Fless, G. M., Scanu, A. M., and Lawn, R. M.,** cDNA sequence of human apolipoprotein (a) is homologous to plasminogen, *Nature,* 330, 132, 1987.
53. **Frank, S. L., Klisak, J., Sparkes, R. S., Mohandas, T., Tomlinson, J. E., McLean, J. W., Lawn, R. M., and Lusis, A. J.,** Apolipoprotein (a) gene resides on human chromosome 6q26-27 in close proximity to the homologous gene for plasminogen, *Hum. Genet.,* 79, 352, 1988.
54. **Utermann, G. and Weber, G.,** Protein composition of Lp(a) lipoprotein from human plasma, *FEBS Lett.,* 154, 357, 1982.
55. **Rhoads, G. G., Dahlen, G., Berg, K., Morton, N. E., and Dannenberg, A. L.,** Lp(a) lipoprotein as a risk factor for myocardial infarction, *JAMA,* 256, 2540, 1986.
56. **Dahlen, G. H., Guyton, J. R., Attar, M., Farmer, J. A., Kautz, J. A., and Gotto, A. M.,** Association of levels of lipoprotein Lp(a), plasma lipids and other proteins with coronary artery disease documented by angiography, *Circulation,* 74, 758, 1986.

57. **Skretting, G., Blomhoff, J. P., Solheim, J., and Prydz, H.,** The genetic defect of the original Norwegian lecithin: cholesterol acyltransferase deficiency families, *FEBS Lett.*, 309, 307, 1992.
58. **O'Donnell, K. A. and Lusis, A. J.,** Genetic evidence that the multiple apolipoprotien A-I isoforms are encoded by a common structural gene, *B.B.R.C.*, 114, 275, 1983.
59. **Sorci-Thomas, M., Prack, M. M., Dashti, N., Johnson, F., Rudel, L. L., and Williams, D. L.,** Apolipoprotein (apo) A-I production and mRNA abundance explain plasma A-I and high density lipoprotein differences between two nonhuman primate species with high and low susceptibilities to diet-induced hypercholesterolemia, *J. Biol. Chem.*, 263, 5183, 1988.
60. **Sorci-Thomas, M. and Kearns, M. W.,** Transcriptional regulation of the apolipoprotein A-I gene. Species specific expression correlates with rates of gene transcription, *J. Biol. Chem.*, 266, 18045, 1991.
61. **Blangero, J., McCluer, J. W., Kammerer, C. M., Mott, G. E., Dyer, T. D., and McGill, H. C.,** Genetic analysis of apolipoprotein A-I in two dietary environments, *Hum. Genet.*, 47, 414, 1990.
62. **Rubin, E. M., Ishida, B. Y., Clift, S. M., and Krauss, R. M.,** Expression of human apolipoprotein A-I in transgenic mice results in reduced plasma levels of murine apolipoprotein A-I and the appearance of two new high density lipoprotein subclasses, *Proc. Natl. Acad. Sci. U.S.A.*, 88, 434, 1991.
63. **Rubin, E. M., Krauss, R. M., Spangler, E. A., Verstuyft, J. G., and Clift, S. M.,** Inhibition of early atherogenesis in transgenic mice by human apolipoprotein A-I, *Nature*, 353, 265, 1991.
64. **Paigen, B., Mitchell, D., Reue, K., Morrow, A., Lusis, A. J., and LeBoeuf, R. C.,** Ath-I, a gene determining atherosclerosis susceptibility and high density lipoprotein levels in mice, *Proc. Natl. Acad. Sci. U.S.A.*, 84, 3763, 1987.
65. **Kushwaha, R. S., Reardon, C. A., Lewis, D. S., Qi, Y., Rice, K. S., Getz, G. S., Carey, K. D., and McGill, H. C.,** Effects of dietary lipids on plasma activity and hepatic mRNA levels of cholesterol ester transfer protein in high and low responding baboons, *Metabolism*, 43, 1006, 1994.
66. **Quinet, E., Tall, A., Ramakrishnan, R., and Rudel, L.,** Plasma lipid transfer protein as a determinant of the atherogenicity of monkey plasma lipoproteins, *J. Clin. Invest.*, 87, 1559, 1991.
67. **Strobl, W., Gorder, N. L., Fienup, G. A., Lin-Lee, Y. C., Gotto, A. M., and Patsch, W.,** Effect of sucrose diet on apolipoprotein synthesis in rat liver. Increase in apo E gene transcription, *J. Biol. Chem.*, 264, 1190, 1989.
68. **Clarke, S. D. and Abraham, S.,** Gene expression: nutrient control of pre and post transcriptional events, *FASEB J.*, 6, 3146, 1992.
69. **Smith, S.,** The animal fatty acid synthase: one gene, one polypeptide, seven enzymes, *FASEB J.*, 8, 1248, 1994.
70. **Dreon, D. M., Fernstrom, H. A., Miller, B., and Krauss, R. M.,** Low density lipoprotein subclass patterns and lipoprotein response to a reduced-fat diet in men, *FASEB J.*, 8, 121, 1994.
71. **Guichard, C., Dugail, I., LeLiepore, X., and Lavau, M.,** Genetic regulation of fatty acid synthetase expression in adipose tissue: over transcription of the gene in genetically obese rats, *J. Lipid Res.*, 33, 679, 1992.
72. **Williams, S. C., Bruckheimer, S. M., Lusis, A. J., LeBoeuf, R. C., and Kinniburgh, A. J.,** Mouse apolipoprotein A-IV gene: nucleotide sequence and induction by a high-lipid diet, *Mol. Cell. Biol.*, 6, 3807, 1986.
73. **Reue, K., Purcell-Huyorh, D. A., Leete, T. H., Doolittle, M. H., Durstenfeld, A., and Lusis, A. J.,** Genetic variation in mouse apolipoprotein A-IV expression is determined pre and post transcriptionally, *J. Lipid Res.*, 34, 893, 1993.

74. **Chen, M., Breslow, J. L., Li, W., and Leff, T.,** Transcriptional regulation of the apo C-III gene by insulin in diabetic mice: correlation with changes in plasma triglyceride levels, *J. Lipid Res.*, 35, 1918, 1994.
75. **Schonfeld, G.,** Dietary treatment of hyperlipidemia, *Clin. Chem.*, 34, B111, 1988.
76. **Muldoon, M. F., Manuck, S. B., and Matthews, K. A.,** Lowering of cholesterol concentrations and mortality: a quantitative review of primary prevention trials, *Brit. Med. J.*, 301, 309, 1990.

Chapter 7

DIET–GENE INTERACTIONS IN THE REGULATION OF LIPOGENESIS

Nobuko Iritani

CONTENTS

INTRODUCTION

It is known that triacylglycerol synthesis is limited by the activities of the enzymes that catalyze fatty acid synthesis. The first step in the pathway of long-chain fatty acid biosynthesis is mediated by acetyl-CoA carboxylase. This enzyme catalyzes the carboxylation of acetyl-CoA to malonyl-CoA. The second step in fatty acid synthesis is the conversion of acetyl-CoA and malonyl-CoA to long-chain fatty acids by the multifunctional enzyme fatty acid synthase complex in the presence of NADPH. This NADPH is provided when reactions catalyzed by malic enzyme, glucose-6-phosphate dehydrogenase, and 6-phosphogluconate dehydrogenase are active. ATP citrate-lyase catalyzes the

0-8493-9485-6/96/$0.00+$.50

formation of acetyl-CoA and oxaloacetate from citrate and CoA with a concomitant hydrolysis of ATP to ADP and phosphate. The product, acetyl-CoA serves as the substrate for several important biosynthetic pathways, including lipogenesis. Thus, ATP citrate-lyase is considered to be one of the lipogenic enzymes. It is well known that the induction of hepatic lipogenic enzymes is increased by feeding a low fat or fat-free/high carbohydrate diet and decreased by fasting, by feeding polyunsaturated fats, or diabetes mellitus. One of the major organs for triacylglycerol synthesis is the liver. A number of studies on the regulation of lipogenic enzyme levels have already been done and reviewed.[1-7] Moreover, the regulation of lipogenic enzyme gene expression has recently been investigated.[8-38] We have investigated the nutritional and hormonal regulation of the gene expression of all the lipogenic enzymes, acetyl-CoA carboxylase, fatty acid synthase, ATP citrate-lyase, malic enzyme, and glucose-6-phosphate dehydrogenase in rat liver, using appropriate cDNAs for these enzymes.[15-19] The purpose of this chapter is to review these studies on the regulation of gene expression.

TIME COURSE FOR LIPOGENIC ENZYME GENE EXPRESSION AFTER REFEEDING A HIGH-CARBOHYDRATE DIET TO FASTED RATS

The transcriptional rates, mRNA concentrations, and activities of lipogenic enzymes in rat liver were markedly decreased by fasting and were greatly increased by refeeding a fat-free/high-carbohydrate diet. The time course of the changes in these parameters after refeeding did not differ greatly among the enzymes, although the maxima for transcription rate and mRNA concentration of acetyl-CoA carboxylase occurred earlier than for the other enzymes (Figure 1).[15-19] In all the experiments described below, immunochemical titrations demonstrated that changes in the activity of any given enzyme were accompanied by proportional changes in the quantity of the enzyme protein. The transcription rates began to increase within 1 h of refeeding and reached maximum levels after 1 to 4 h. The mRNA concentrations began to increase after 4 h and reached their maxima after 8 to 16 h. Activities of acetyl-CoA carboxylase, fatty acid synthase, and ATP citrate-lyase reached maximum levels by 48 h after refeeding, and those of malic enzyme and glucose-6-phosphate dehydrogenase after 72 h.

After feeding the rats the high-carbohydrate diet, the mRNA concentrations of all the lipogenic enzymes studied increased to a greater extent than the respective transcription rates. The magnitudes of enzyme induction, with the exception of fatty acid synthase and ATP citrate-lyase, roughly paralleled the mRNA concentrations. These results suggest that the high-carbohydrate diet stimulated primarily transcription and resulted in an increase in mRNA stability. The transcription rates and mRNA concentrations of fatty acid synthase and ATP citrate-lyase were markedly increased in comparison with those of the other enzymes, whereas the magnitudes of enzyme induction were similar for

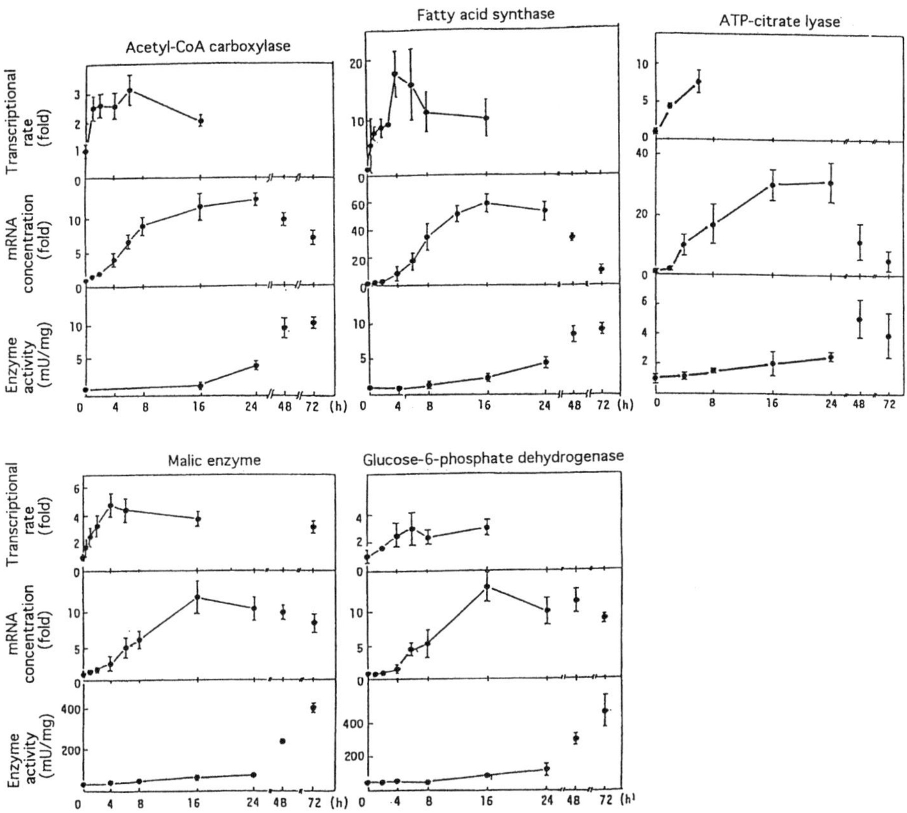

FIGURE 1. Time course of changes in transcriptional rates, mRNA concentrations and induction of hepatic lipogenic enzymes after feeding the high-carbohydrate diet to fasted rats. After refeeding the high-carbohydrate diet to fasted rats, transcriptional rates, mRNA concentrations, and activities of hepatic lipogenic enzymes were followed over time. The high-carbohydrate diet contained 67% sucrose, 18% casein, 9.9% cellulose, 5% salt mixture, 0.1% choline chloride and vitamins.[39] The animals were killed at the times indicated after refeeding. The enzyme activities in the supernatant of the liver homogenate are shown as mU/mg protein, where 1 mU is the amount catalyzing the formation of 1 nmol product/min at 37°C for acetyl-CoA carboxylase, fatty acid synthase, and ATP citrate-lyase, and the utilization of NADP for malic enzyme and glucose-6-phosphate dehydrogenase. Other data are normalized to the values of fasted rats (at zero time). Mean ± SD (n = 3–9). (Taken from Katsurada, A., Iritani, N., Fukuda, H., et al.[15-19] With permission.)

all the lipogenic enzymes. As the mRNA concentrations of acetyl-CoA carboxylase, ATP citrate-lyase, and fatty acid synthase decreased greatly after attaining each maximum level, the mRNA concentrations appeared to be unstable. The levels of all the lipogenic enzyme mRNA species, 72 h after refeeding the high-carbohydrate diet, were five to ten times higher than each respective fasting level.

From the time course of mRNA induction, the half lives of the lipogenic enzyme mRNA species were calculated to be 4.5 to 6 h, whereas the lag times for the mRNA of malic enzyme and glucose-6-phosphate dehydrogenase were longer than those of the other enzymes. The enzyme half lives of acetyl-CoA carboxylase, fatty acid synthase, and ATP citrate-lyase were 20 to 23 h and those of malic enzyme and glucose-6-phosphate dehydrogenase were 26 to 30 h. In our previous study,[40] in which rats adapted to a commercial stock diet were given a fat-free diet and vice versa, the half lives of the hepatic lipogenic enzymes exceeded 2 d. Another group reported that the half life of hepatic acetyl-CoA carboxylase in fed rats at steady state was 55 to 59 h.[41] The half-lives of lipogenic enzymes and their mRNA species could thus be altered by changes in nutritional conditions. Differences in half lives have been reported for different species. For example, Goodridge et al.[30-33] reported half lives for avian enzymes. They estimated the half life of hepatic fatty acid synthase mRNA to be 4 to 6 h for fed ducklings and about 3 h in fasted ducklings,[30] whereas the half life of hepatic malic enzyme mRNA in fed ducklings was 3 h.[33]

EFFECTS OF DIET INTAKE QUANTITY ON LIPOGENIC ENZYME GENE EXPRESSION

The effects of the quantity of the food consumed using a high carbohydrate diet on the maximal transcriptional rates, mRNA concentrations, and induction of lipogenic enzymes were investigated. Fasted rats were fed different amounts of a high-carbohydrate diet and killed after 6, 16, and 72 h.[21] After 6 h, the transcriptional rates of acetyl-CoA carboxylase and malic enzyme reached maximal levels in rats fed 25% of their usual ad libitum dietary intake. Maximal transcription rates of fatty acid synthase and glucose-6-phosphate dehydrogenase were achieved in rats fed 50% of their ad libitum intake (Figure 2).[21] However, the mRNA concentrations of the lipogenic enzymes continued to increase as the dietary intake increased. It is remarkable that the transcription rates of the lipogenic enzymes generally reached maximum levels with a low dietary intake, whereas the mRNA concentrations continued to increase greatly with the intake of larger quantities. This suggests, therefore, that an increase of dietary intake initiates transcription and increases the stability of mRNA.

Sixteen hours after refeeding different quantities of the high-carbohydrate diet, the mRNA concentrations of all the enzymes increased sigmoidally with the diet quantity and reached maxima 15 to 30 times higher than the initial levels, with the exception of fatty acid synthase. This enzyme mRNA reached extremely high levels (a 110-fold increase) at this time point. After refeeding the high-carbohydrate diet for 3 d, the induction of all lipogenic enzymes reached maximum. The enzyme induction increased sigmoidally with dietary intake and reached maxima about 20 times higher than the initial levels (fasted), with the exception of ATP citrate-lyase which reached maximum of only 5 times that of the initial values (Figure 3).

After injecting rats with tritiated water, the incorporation of radioactivity into hepatic fatty acids also increased sigmoidally with dietary intake (Figure 3).

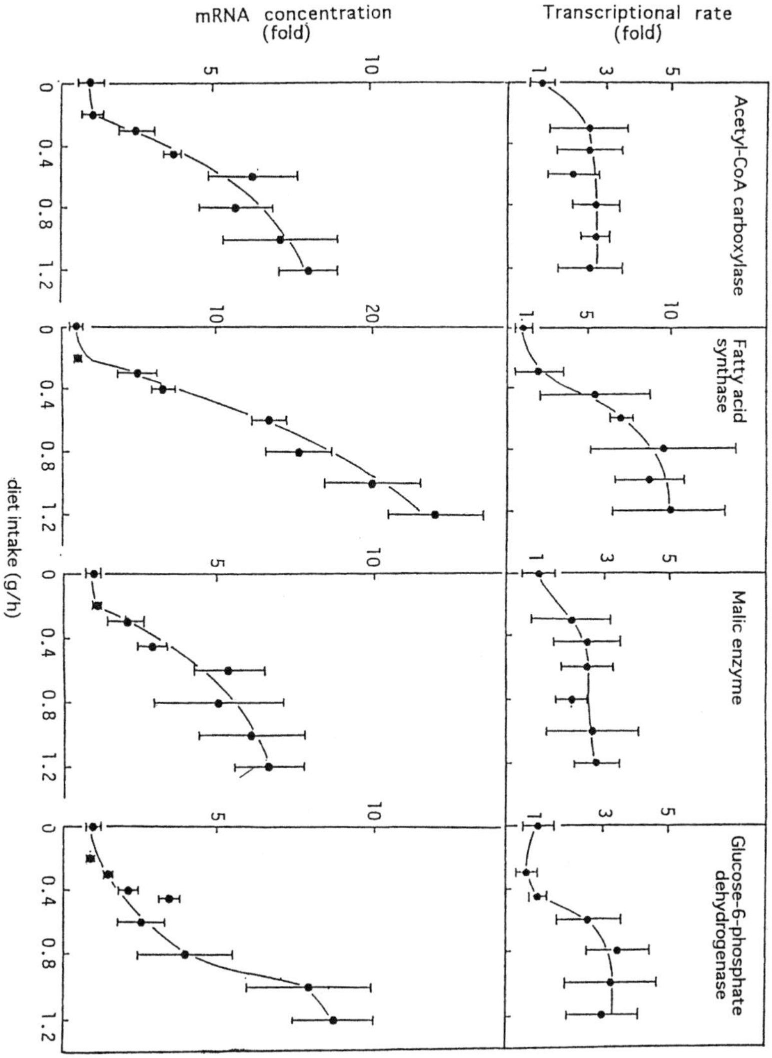

FIGURE 2. Effects of intake quantity of the high carbohydrate diet on transcriptional rates and mRNA concentrations of lipogenic enzymes in rat livers after feeding for 6 h. The fasted rats were fed the high-carbohydrate diet (Figure 1) of 0.2 to 1.2 g per 100 g body mass every 1 h for 6 h, and then killed. The transcriptional rates and mRNA concentrations are normalized to the values of fasted rats (at 0-g intake). Mean ± SD (n = 3). (Taken from Iritani, N., et al.[21] With permission.)

From these results, it appeared that fatty acid synthesis was not increased in rats fed less than 50% of the ad libitum intake, but began to increase in those fed 70% of the ad libitum intake and increased greatly when the amount supplied exceeded this. However, lipogenic enzyme induction began to increase slightly in rats fed 30% of the ad libitum intake, and was increased about tenfold in rats

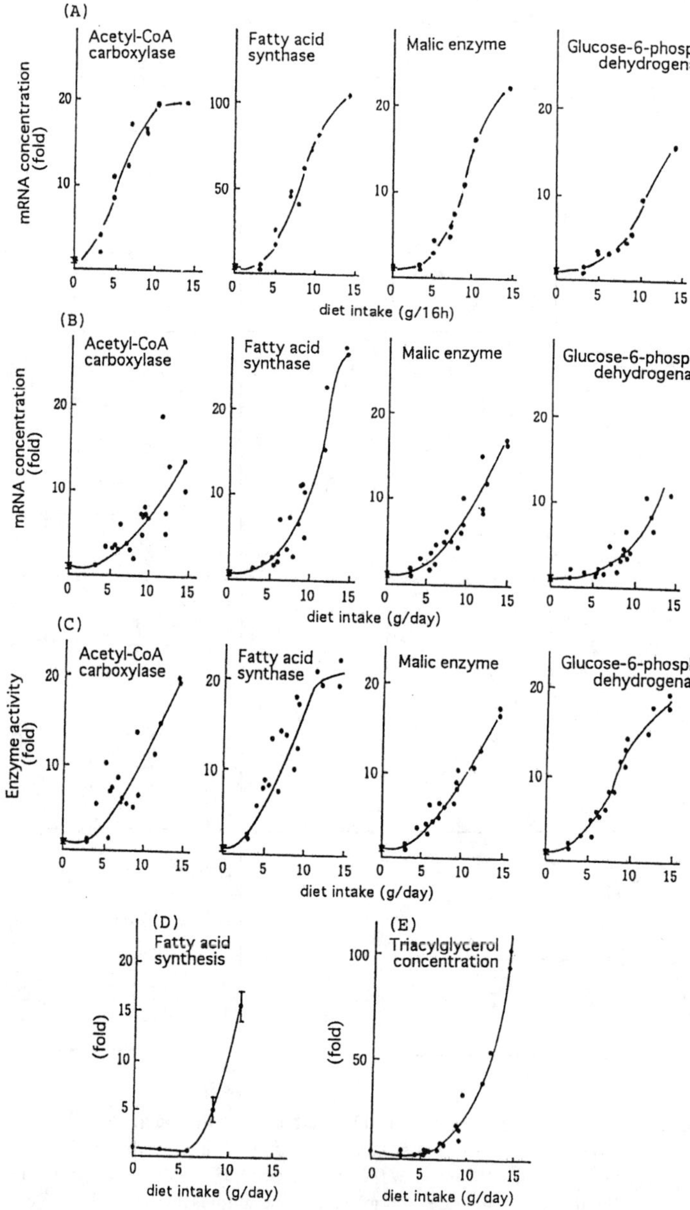

FIGURE 3. Effects of intake quantity of the high-carbohydrate diet on lipogenic enzyme induction, fatty acid synthesis, and triacylglycerol levels in rat livers. The fasted rats were fed 3 to 14 g per 100 g/d of the high-carbohydrate diet and killed after 16 h (A) or 3 d (B, C, D, E). The rats were injected intraperitoneally with 7.4 MBq tritiated water per 100 g body mass, then killed after 1 h. The radioactivities of the fatty acid fraction in the liver were measured. A: mRNA concentrations after 16 h; B: mRNA concentrations after 3 d; C: enzyme induction; D: fatty acid synthesis; E: triacylglycerol concentrations. The data are normalized to the values of fasted rats. (Taken from Iritani, N., et al.[21] With permission.)

fed 70% of the ad libitum intake. Therefore, in addition to gene expression of lipogenic enzymes, other factors, such as hormone levels and substrate and effecter concentrations, may affect the rate of fatty acid synthesis in this situation. The concentrations of acetyl-CoA, malonyl-CoA, L-malate and glucose-6-phosphate were decreased to half the normal level by fasting.[42] The concentrations of citrate, an allosteric activator of acetyl-CoA carboxylase,[1] also decreased similarly with fasting.[42] Thus, the concentrations of substrates and/or effectors are such that fatty acid synthesis is not stimulated in rats fed less than 50% of their ad libitum intake, even though the lipogenic enzymes are induced. It appeared that the earlier the step in the fatty acid synthesis, the lesser the quantity of diet required for its stimulation. Moreover, the triacylglycerol level increased with increasing dietary intake, and the change roughly paralleled that in fatty acid synthesis.

EFFECTS OF DIETARY NUTRIENTS ON LIPOGENIC ENZYME GENE EXPRESSION

The effects of dietary nutrients on the transcriptional rates, mRNA concentrations, and induction of lipogenic enzymes were compared at their respective maximum levels, 6, 16, and 72 h after refeeding equal energy diets of either carbohydrate/protein, carbohydrate, protein or carbohydrate/protein/fat to fasted rats.[15-19] The transcriptional rates and mRNA concentrations of acetyl-CoA carboxylase and ATP citrate-lyase with the protein-free carbohydrate diet were similar to those with the carbohydrate/protein diet, and similar findings were also obtained even for the enzyme inductions (Figure 4). It appears, therefore, that the induction of acetyl-CoA carboxylase and ATP citrate-lyase are strongly dependent on dietary carbohydrate and do not require protein. Dietary carbohydrate increased the transcriptional rate, mRNA concentration, and induction of fatty acid synthase to a greater extent than protein.[17] However, both carbohydrate and protein were required in order to achieve a substantial increase.[42]

The transcription rate and mRNA concentration of malic enzyme, after feeding with carbohydrate only, were increased to levels similar to those in rats fed the carbohydrate/protein diet, whereas in rats fed carbohydrate alone, induction of the enzyme was increased to only 60% of the level obtained with a mixture of carbohydrate and protein diet.[19] Thus, it appears that protein feeding does not contribute in the transcriptional step or to mRNA stability, but rather to an increase in the translation of magic enzyme mRNA. Conversely, the transcriptional rate of glucose-6-phosphate dehydrogenase tended to be increased after feeding with protein rather than carbohydrate.[18] The enzyme induction and mRNA concentration were also increased more by protein than by carbohydrate. Thus, it appears that magic enzyme induction is carbohydrate-rather than protein-dependent, and that the converse applies to induction of glucose-6-phosphate dehydrogenase.

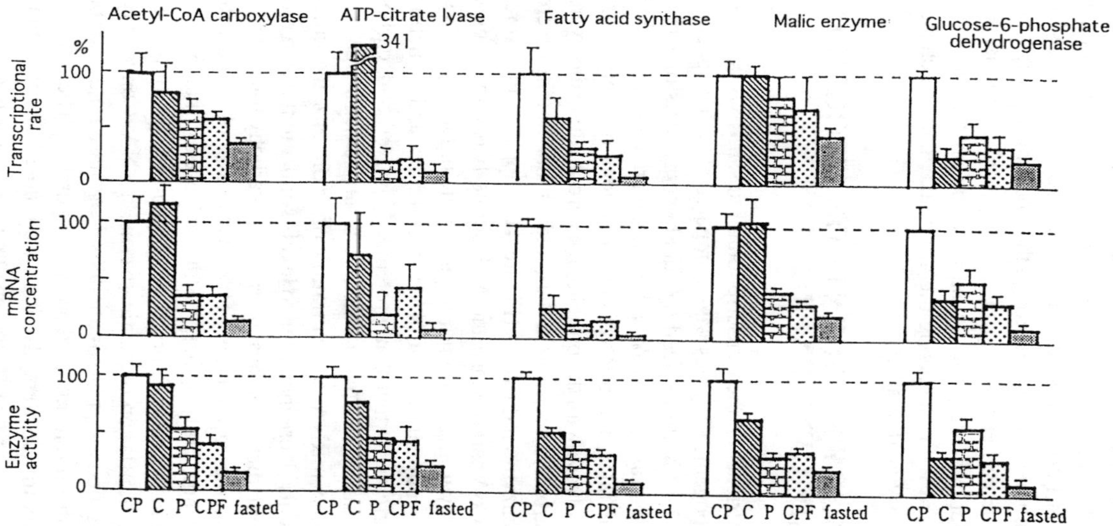

FIGURE 4. Effects of dietary nutrients on transcriptional rates, mRNA concentrations, and induction of lipogenic enzymes in rat livers. Rats were fasted for 2 d, then refed with diets containing the same calories of either a 67% carbohydrate/18% protein (CP), a 85% carbohydrate (C), a 85% protein (P), or a 57% carbohydrate/18% protein/10% fat (CPF) diet. The carbohydrate/protein diet was the same as shown in the legend of Figure 1. For the CPF diet, the amount of 10% sucrose (by mass) was substituted by corn oil, which contained 50.2% linoleic acid. The transcriptional rates, mRNA concentrations, and enzyme activities were measured in each maximum level at 6, 16, and 72 h, respectively, after the refeeding. The results are normalized to the values for the carbohydrate/protein group. The enzyme activities of acetyl-CoA carboxylase, ATP citrate-lyase, fatty acid synthase, malic enzyme, and glucose-6-phosphate dehydrogenase in the carbohydrate/protein group were 9.50 ± 1.14, 5.99 ± 0.79, 9.86 ± 0.79, 350 ± 38.5 and 392 ± 48.1 mU/mg protein of the supernatant of liver homogenate at 37°C, respectively. Mean ± SD (n = 6–9). (Taken from Katsurada, A., Iritani, N., Fukuda, H., et al.[15-19] With permission.)

In the portal vein, the plasma glucagon levels in rats fed carbohydrate only or carbohydrate and protein did not differ markedly, although the levels were notably higher in animals fed a carbohydrate-free protein diet than in those fed a carbohydrate/protein diet.[42] Unger et al.[43,44] also reported that dietary protein raised the level of plasma glucagon. Hepatic cyclic AMP levels tended to be higher in the protein-fed group than in the other dietary groups[42] and were reported to increase after glucagon injection.[45] Glucagon and cyclic AMP were demonstrated to prevent the induction of rat liver glucose-6-phosphate dehydrogenase.[46,47] Therefore, the increase in glucagon and cyclic AMP levels with the protein diet may have prevented enzyme induction.

In rats adapted to a carbohydrate/protein diet, feeding corn oil, rich in n-6 polyunsaturated fatty acid (linoleic acid), reduced the transcriptional rates, mRNA concentrations, and induction of lipogenic enzymes to 30 to 40% after 6 h, 16 h, and 3 d, respectively.[15-19] Dietary corn oil primarily decreased the transcriptional rate. Similar results were obtained after feeding fish oil or perilla oil, which are rich in n-3 polyunsaturated fatty acids (eicosapentaenoic acid and docosahexaenoic acid in the former, and linolenic acid in the latter).[26,27,48] It appears that the fatty acids must be at least 18 carbons in length and contain at least 2 conjugated double bonds located at the 9 and 12 position[11] in order to actively inhibit fatty acid synthesis. The n-9 polyunsaturated fatty acids did not suppress fatty acid synthesis.

Furthermore, the fatty acid composition of hepatic nuclear phospholipids was altered by manipulation of dietary fatty acid species.[49] In rats fed corn or fish oil, the respective proportions of n-6 or n-3 polyunsaturated fatty acids among hepatic nuclear phospholipids were increased in a few hours. Moreover, in rats injected with ^{3}H-labeled exogenous fatty acids, the radioactivity was incorporated into hepatic nuclear triacylglycerols and phospholipids within 1 h after injection.[50] Further support for these results was provided by an autoradiographic study.[50] Therefore, dietary fatty acids appear to be incorporated rapidly into nuclear fatty acids and thus may be involved in nuclear and cytosolic regulation of lipogenic enzyme induction. Although the transcriptional rates were decreased by feeding with polyunsaturated fats, the mechanism of this effect has not been elucidated.

However, it has recently been reported that the cis-linked polyunsaturated fatty acid response elements are localized in the promoter region of the L-pyruvate kinase gene and also in that of S14, which has been used as a model for examining the effects of polyunsaturated fatty acids on hepatic lipogenic enzyme gene expression.[51,52] Moreover, a carbohydrate response element, sharing a segment with 9 out of 10 identity, has already been found in the promoters of the L-pyruvate kinase and S14 genes.[53-55]

On the other hand, we have found that the expression of lipogenic enzyme genes was not suppressed by dietary polyunsaturated fatty acids in genetically obese Wistar fatty rats (with hyperinsulinemia and noninsulin-dependent diabetes mellitus), whereas the expression was markedly suppressed in lean rats.[56] Binding of insulin to receptors in the liver was not reduced by polyunsaturated fatty acids in fatty rats, but was reduced in lean rats. Insulin binding to receptors was decreased in the fatty rats and the sensitivity of insulin receptors to polyunsaturated fatty acids appeared to be low, probably due to down-regulation induced by hyperinsulinemia. Thus, defects in the polyunsaturated fatty acid-mediated suppression of lipogenic enzyme gene expression in the fatty rats could be ascribed to a decrease of the function of insulin receptors.

Moreover, we have investigated the effects of dietary polyunsaturated fat on insulin-dependent gene expression of lipogenic enzymes, and a possible mechanism for polyunsaturated fatty acid-mediated suppression of the gene expression.[57] When streptozotocin diabetic rats were injected with insulin, the insulin dose-dependent induction of lipogenic enzyme mRNAs was markedly reduced with increasing dietary corn oil in a dose-dependent fashion. The polyunsaturated fatty acid-mediated suppression of the mRNA concentrations was partially restored by treatment with pioglitazone, a candidate drug for increasing insulin receptor phosphorylation. In these animals, insulin binding to hepatic receptors, receptor autophosphorylation, and kinase activity were lowered by corn oil feeding. However, the polyunsaturated fatty acid-mediated suppression of insulin binding was somewhat restored by pioglitazone, and the suppression of insulin receptor phosphorylation was significantly restored. Thus, it is suggested that the polyunsaturated fatty acid-mediated suppression of lipogenic enzyme gene expression can be involved in the function of insulin receptors. Moreover, polyunsaturated fatty acids suppressed the lipogenic enzyme gene expression stimulated by insulin.

EFFECTS OF INSULIN ON LIPOGENIC ENZYME GENE EXPRESSION

The transcription rates, mRNA concentrations, and activities (adjusted in proportion to enzyme levels) of lipogenic enzymes in diabetic rats adapted to a high-glucose diet were generally 30 to 40%, 20 to 30%, and 10 to 20% of the respective levels in normal rats, but were restored to normal 1 to 6 h, 16 h, and 2 to 3 d after insulin treatment, respectively (Figure 5).[15-19] The insulin-induced restoration of the transcriptional rates of these lipogenic enzymes in diabetic rats was reduced markedly by injection of cycloheximide. Therefore, *de novo* enzyme protein synthesis appeared to be required for the insulin-induced enhancement of transcription.

Feeding a high-carbohydrate diet to normal fasted animals induced a marked overshoot of transcription and mRNA concentration, particularly those of fatty acid synthase. However, treatment of diabetic rats with insulin produced a rapid and marked increase in the transcriptional rat fatty acid synthase, although the mRNA concentration was restored only to the normal level. These results suggest that insulin acts by stimulating transcription rather than by enhancing mRNA stability, whereas carbohydrate stimulates transcription and in particular, increases mRNA stability. Recently, the novel cis-acting DNA

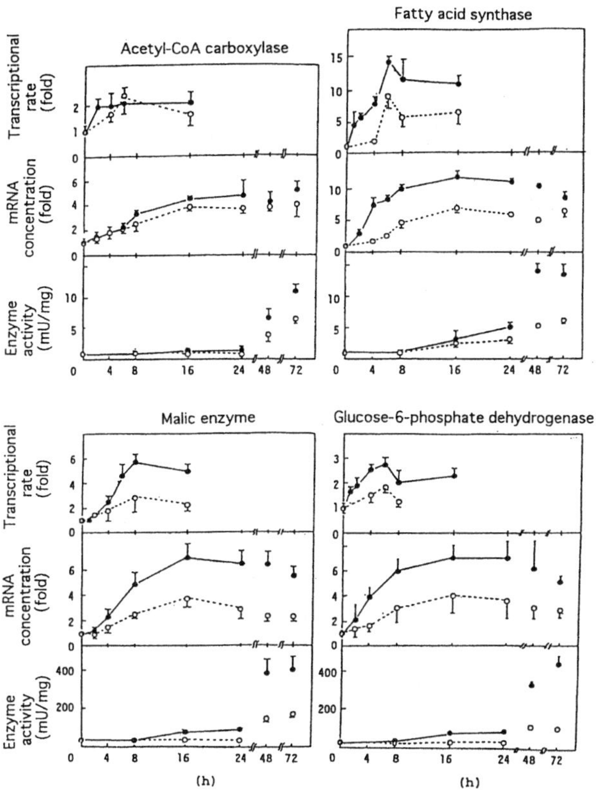

FIGURE 5. Changes on transcriptional rated mRNA concentrations and induction of lipogenic enzymes in diabetic rat livers after insulin treatment or fructose feeding. For insulin treatment, diabetic rats fed on a glucose diet were injected with 0.5 I.U. Actrapid insulin intraperitoneally and 1.5 I.U. Lente insulin subcutaneously (solid line). For continuous insulin treatment (to measure the enzyme activities), the diabetic rats were subcutaneously injected with 2.5 I.U. Lente insulin/day (1 I.U. in the morning and 1.5 I.U. in the evening). For fructose feeding, the diabetic animals were given 200 mg fructose orally and shifted to a fructose diet (dashed line). For the glucose or fructose diet, sucrose in the diet shown in Figure 1 was substituted by glucose or fructose. The animals were killed at the times indicated. The transcriptional rate and mRNA concentration are normalized to the values of diabetic rats. Each value is the mean ± SD (n = 6–9). (Taken from Katsurada, A., Iritani, N., Fukuda, H., et al.[15-19] With permission.)

sequences responsible for insulin regulation of the fatty acid synthase gene have been found in rat liver.[58,59] In an *in vivo* study, however, the expression of lipogenic enzyme genes was not increased by insulin in the liver, unless the animal was given a high-carbohydrate diet. Moreover, gene expression in cultured hepatocytes was not increased by insulin in the absence of glucose.[60] Therefore, it appears that glycolytic metabolites primarily increase transcription, although insulin may have a permissive action in amplifying the transcription through the cis-acting element.

EFFECTS OF FRUCTOSE-FEEDING ON LIPOGENIC ENZYME GENE EXPRESSION

A fructose diet, even in diabetic rats, increases the levels of glycolytic and lipogenic enzymes, compared with a glucose diet.[15-19,38,61-65] With respect to glycolysis, both glucose and fructose have been shown to stimulate expression of the hepatic L-type pyruvate kinase gene at the transcription step.[66] A carbohydrate-responsive element has been identified in the promoter region of the L-pyruvate kinase gene,[53,66,67] and the enhancer unit is responsive not only to dietary glucose, but also to dietary fructose in transgenic mice, suggesting that glucose and fructose share a common action of transcriptional stimulation.[66,67]

In diabetic animals, it is postulated that the concentrations of glycolytic metabolites, which are increased as a consequence of the primary action of insulin, which induces glucokinase,[68] are decreased. Even in a diabetic state, however, the level of fructokinase is not reduced markedly, despite the low level of hepatic glucokinase,[69-71] and fructose is metabolized to yield an appreciable level of glycolytic intermediates. Therefore, we can investigate the effects of glycolytic metabolites on gene expression in the presence of a low insulin level after feeding fructose to diabetic rats.

In our previous studies,[15-19] after diabetic animals adapted to a glucose diet were switched to a high-fructose diet, the time course of the changes in transcriptional rates, mRNA concentrations, and enzyme induction of the lipogenic enzymes were followed and compared with those induced by insulin treatment (Figure 5). After feeding fructose to diabetic rats, the transcriptional rates and mRNA concentrations of acetyl-CoA carboxylase and ATP citrate-lyase were increased to levels similar to, or even higher than those attained with insulin treatment,[15,16] suggesting that these increases were induced by glycolytic metabolites. This is in accord with the results of the nutrition study described above, where acetyl-CoA carboxylase and ATP citrate-lyase were the most carbohydrate-dependent of the lipogenic enzymes. However, it was remarkable that the increase in enzyme induction resulting from fructose feeding was only 60% of that obtained with insulin. None of the increases in the magnitude of lipogenic enzyme induction were as great as that of mRNA concentration. Therefore, insulin appears to be required for both translation and transcription of lipogenic enzymes.

EFFECTS OF TRIIODOTHYRONINE ON LIPOGENIC ENZYME GENE EXPRESSION

In normal rats, triiodothyronine treatment increased the synthesis of malic enzyme markedly, primarily at the transcription level, and increased the synthesis of acetyl-CoA carboxylase and fatty acid synthase slightly but did not increase the synthesis of glucose-6-phosphate dehydrogenase and ATP citrate-lyase.[15-19] Triiodothyronine has a marked effect on malic enzyme transcription, as reported elsewhere.[13,19,28,33,36] The triiodothyronine response element has been found in the 5′ flanking region of the malic enzyme gene.[28]

The effect of triiodothyronine treatment on lipogenic enzyme gene expression was greater in diabetic than in normal rats. Treatment of diabetic animals with triiodothyronine increased the expression of the acetyl-CoA carboxylase, fatty acid synthase, and malic enzyme genes greatly, but the expression of the glucose-6-phosphate dehydrogenase and ATP citrate-lyase genes increased slightly. Diabetes is associated with hypothyroidism.[72] The plasma levels of triiodothyronine in diabetic rats are reported to be much lower than in normal rats.[73] In earlier studies, we found that the concentrations of hepatic glucose-6-phosphate and fructose-6-phosphate, which were reduced in diabetic rats, were increased by treatment with insulin or triiodothyronine.[17,64] Therefore, the effects of triiodothyronine treatment on gene expression in diabetic animals appear to be due, at least partly, to the increased glycolytic metabolites.

COORDINATED REGULATION OF GLYCOLYTIC AND LIPOGENIC ENZYME GENE EXPRESSION

The regulation of gene expression of glycolytic enzymes in rat liver was compared with the regulation of lipogenic enzymes during nutritional manipulation.[74] After refeeding a high carbohydrate diet to fasted rats, the mRNA and enzyme inductions of glucokinase reached maximum levels in 2 h and 16 h, respectively, whereas those of L-pyruvate kinase reached maximum levels in 16 and 48 h, respectively, showing similar time courses to those of lipogenic enzymes. The mRNA and enzyme inductions of L-pyruvate kinase increased sigmoidally with the quantity of food consumed and this was similar to the responses of the lipogenic enzymes. However, the mRNA and enzyme inductions of glucokinase were increased in rats fed only a small quantity of food. By feeding rats a carbohydrate diet without protein, the increases in the mRNA and enzyme inductions of L-pyruvate kinase and glucokinase were comparable to the levels obtained when a carbohydrate/protein diet was fed. Thus, it appears that protein feeding is not necessary for the induction of glucokinase and L-pyruvate kinase in normal rats. The induction of L-pyruvate kinase mRNA by treatment of diabetic rats with insulin or by feeding a carbohydrate diet to fasted animals was markedly reduced by dietary polyunsaturated fatty acids, whereas the induction of glucokinase mRNA was only slightly reduced by polyunsaturated fatty acids. Jump et al.[75] have reported the coordinated regulation of glycolytic and lipogenic gene expression by polyunsaturated fatty acids. Moreover, the cis-linked polyunsaturated fatty acid response elements are reported to be localized in the promoter region of the L-pyruvate kinase gene.[51] It appears that L-pyruvate kinase gene expression is regulated by dietary nutrients in a manner that is coordinated with the lipogenic enzyme genes. However, the glucokinase gene expression is rapidly induced even when a small amount of carbohydrate is consumed.

SUMMARY

After refeeding a high-carbohydrate diet to fasted rats, the transcriptional rates, mRNA concentrations and induction of lipogenic enzymes reached maximum levels in 14 h, 8 to 16 h, and 2 to 3 d, respectively. The transcriptional rates generally reached maximum with a low dietary intake, whereas the mRNA concentrations continued to increase with increasing dietary intake. Therefore, increased dietary intake appears to initiate transcription and increase the mRNA concentration. The transcription rates, mRNA concentrations, and even the induction of acetyl-CoA carboxylase and ATP citrate-lyase activity were markedly increased to the same extent by feeding rats with carbohydrate alone, or with both carbohydrate and protein. The transcription and mRNA levels of malic enzyme were also increased by feeding with carbohydrate alone, although feeding with both carbohydrate and protein was required for enzyme induction. The gene expression of fatty acid synthase was increased several-fold by carbohydrate or protein alone, but did not reach that level achieved with the combination of carbohydrate and protein. Glucose-6-phosphate dehydrogenase gene expression required feeding with both carbohydrate and protein, although it was more dependent on protein than on carbohydrate. Thus, the inductions of acetyl-CoA carboxylase and ATP citratelyase are considered to be those most dependent on carbohydrate among the various lipogenic enzymes.

A fructose diet, which increased the levels of hepatic glycolytic metabolites, increased the transcriptional rates and mRNA concentrations of lipogenic enzymes, even in diabetic rats. The transcription appeared to be stimulated primarily by glycolytic metabolites, suggesting that glucose and fructose share a common transcriptional stimulation mechanism. However, unless diabetic rats received insulin treatment, the magnitude of lipogenic enzyme induction resulting from fructose feeding was less than the increase in their mRNA concentrations. Therefore, insulin appears to be involved in translation as well as transcription during lipogenic enzyme induction.

On the other hand, insulin had a common effect on the gene expression of all the lipogenic enzymes, whereas the effect of triiodothyronine was not common. An insulin response element has been identified in the promoter region of the fatty acid synthase gene. However, the gene expression of lipogenic enzymes in cultured hepatocytes was not increased by insulin in the absence of carbohydrate. Therefore, it appears that glycolytic metabolites primarily initiate transcription, although insulin has a permissive action in amplifying the transcription through the cis-acting element. Polyunsaturated fatty acids suppressed the lipogenic enzyme gene expression stimulated by insulin. The polyunsaturated fatty acid-mediated suppression can be involved in the function of insulin receptors. Very recently, however, the cis-acting responsive elements for carbohydrate and polyunsaturated fatty acids have been elucidated in the L-pyruvate kinase gene, which expression was increased by carbohydrate and suppressed by polyunsaturated fatty acids in a coordinated manner along with lipogenic enzyme genes. These response elements have scarcely been identified in lipogenic enzyme genes. The concept that cis-acting elements of genes for nutritional and hormonal regulation can be involved in the expression is generally accepted. Moreover, the nutrients, carbohydrate, and polyunsaturated fatty acids, themselves can be effectors of gene expression for metabolic regulation. The regulatory genes and the initiation mechanisms responsible for lipogenic enzyme transcription will almost certainly be clarified in the near future.

REFERENCES

1. **Numa, S., Nakanishi, T., Hashimoto, T., Iritani, N., and Okazaki, T.,** Role of acetyl-CoA carboxylase in the control of fatty acid synthesis, *Vitamins and Hormones,* 28, 213, 1970.
2. **Volpe, J. J. and Vagelos, P. R.,** Saturated fatty acid biosynthesis and its regulation, *Ann. Rev. Biochem.,* 42, 21, 1973.
3. **Moss, J. and Lane, M. D.,** The biotin-dependent enzymes, *Adv. Enzymol.,* 35, 321, 1971.
4. **Kim., K.-H.,** Regulation of acetyl-CoA carboxylase, *Curr. Top. Cell. Regul.,* 22, 143, 1983.
5. **Wakil, S. J., Stoops, J. K., and Joshi, V. C.,** Fatty acid synthesis and its regulation, *Ann. Rev. Biochem.,* 52, 537, 1983.
6. **Gibson, D. M., Lyons, R. T., Scott, D. F., and Muto, Y.,** Synthesis and degradation of the lipogenic enzymes of rat liver, *Adv. Enzyme. Regul.,* 10, 187, 1971.
7. **Hardie, D. G.,** The regulation of fatty acid synthesis by reversible phosphorylation of acetyl-CoA carboxylase, in *Molecular Aspects of Cellular Regulation,* Vol. 1, Cohen, P., Ed., Elsevier/North-Holland Biochemical Press, Amsterdam, 1980, 32.
8. **Iritani, N.,** Nutritional and hormonal regulation of lipogenic enzyme gene expression in rat liver, *Eur. J. Biochem.,* 205, 433, 1992.
9. **Clarke, S. D. and Abraham, S.,** Gene expression: nutrient control of pre- and post-transcriptional events, *FASEB J.,* 6, 3146, 1992.
10. **Girard, J., Perdereau, d., Foufelle, F., PripBuus, C., and Ferre, P.,** Regulation of lipogenic enzyme gene expression by nutrients and hormones, *FASEB J.,* 8, 36, 1994.
11. **Clarke, S. D. and Jump, D. B.,** Dietary polyunsaturated fatty acid regulation of gene transcription, *Ann. Rev. Nutr.,* 14, 83, 1994.
12. **Kim, K.-H. and Tae, H.-J.,** Pattern and regulation of acetyl-CoA carboxylase gene expression, *J. Nutr.,* 124, 1273S, 1994.
13. **Goodridge, A. G.,** Dietary regulation of gene expression: enzymes involved in carbohydrate and lipid metabolism, *Ann. Rev. Nutr.,* 7, 157, 1987.
14. **Tao, T.-Y. and Towel, H. C.,** Coordinate regulation of rat liver genes by thyroid hormone and dietary carbohydrate, *Ann. N.Y. Acad. Sci.,* 478, 20, 1986.
15. **Fukuda, H., Katsurada, A., and Iritani, N.,** Effects of nutrients and hormones on gene expression of ATP citrate-lyase in rat liver, *Eur. J. Biochem.,* 209, 217, 1992.
16. **Katsurada, A., Iritani. N., Fukuda, H., Matsumura, Y., Nishimoto, N., Noguchi, T., and Tanaka, T.,** Effects of nutrients and hormones on transcriptional and posttranscriptional regulation of acetyl-CoA carboxylase in rat liver, *Eur. J. Biochem.,* 190, 435, 1990.
17. **Katsurada, A., Iritani, N., Fukuda, H., Matsumura, Y., Nishimoto, N., Noguchi, T., and Tanaka, T.,** Effects of nutrients and hormones on transcriptional and post-transcriptional regulation of fatty acid synthase in rat liver, *Eur. J. Biochem.,* 190, 427, 1990.

18. **Katsurada, A., Iritani, N., Fukuda, H., Matsumura,Y., Nishimoto, N., Noguchi, T., and Tanaka, T.,** Effects of nutrients and insulin on transcriptional and post-transcriptional regulation of glucose-6-phosphate dehydrogenase synthesis in rat liver, *Biochim. Biophys. Acta,* 1006, 104, 1990.
19. **Katsurada, A., Iritani, N., Fukuda, H., Noguchi, T., and Tanaka, T.,** Influence of diet on the transcriptional and post-transcriptional regulation of malic enzyme induction in the rat liver, *Eur. J. Biochem.,* 168, 487, 1987.
20. **Katsurada, A., Iritani, N., Fukuda, H., and Matsumura, Y.,** Effects of insulin and fructose on transcriptional and post-transcriptional regulation of malic enzyme synthesis in diabetic rat liver, *Biochim. Biophys. Acta,* 1004, 103, 1989.
21. **Iritani, N., Nishimoto, N., Katsurada, A., and Fukuda, H.,** Regulation of hepatic lipogenic enzyme gene expression by diet quantity in rat fed a fat-free, high carbohydrate diet, *J. Nutr.,* 122, 28, 1991.
22. **Song, M.-K., Dozin, B., Grieco, D., Rall, J. E., and Nikodem, V. M.,** Transcriptional activation and stabilization of malic enzyme mRNA, *J. Biol. Chem.,* 263, 17970, 1988.
23. **Paulauskis, J. D. and Sul, H. S.,** Hormonal regulation of mouse fatty acid synthase gene transcription in liver, *J. Biol. Chem.,* 264, 574, 1989.
24. **Lopez-Casillas, F. and Kim, K.-H.,** Heterogeneity at the 5′ end of rat acetyl-CoA carboxylase mRNA, *J. Biol. Chem.,* 264, 7176, 1988.
25. **Pape, M. E., Lopez-Casillas, F., and Kim, K.-H.,** Physiological regulation of acetyl-CoA carboxylase gene expression: effects of diet, diabetes, and lactation on acetyl-CoA carboxylase mRNA, *Arch. Biochem. Biophys.,* 267, 104, 1988.
26. **Blake, W. and Clarke, S. D.,** Suppression of rat hepatic fatty acid synthase and S14 gene transcription by dietary polyunsaturated fat, *J. Nutr.,* 120, 1727, 1990.
27. **Clarke, S. D., Armstrong, M. K., and Jump, D. D.,** Dietary polyunsaturated fats uniquely suppress rat liver fatty acid synthase and S14 mRNA content, *J. Nutr.,* 120, 225, 1990.
28. **Petty, K. J., Morioka, H., Mitsuhashi, T., and Nikodem, V. M.,** Thyroid hormone regulation of transcription factors involved in malic enzyme gene expression, *J. Biol. Chem.,* 264, 11483, 1989.
29. **Jump, D. B., Bell, A., and Santiago, V.,** Thyroid hormone and dietary carbohydrate interact to regulate rat liver S14 gene transcription and chromatin structure, *J. Biol. Chem.,* 265, 3474, 1990.
30. **Back, D. W., Goldman, M. J., Fisch, J. E., Ochs, R. S., and Goodridge, A. G.,** The fatty acid synthase gene in avian liver; two mRNAs are expressed and regulated in parallel by feeding, primarily at the level of transcription, *J. Biol. Chem.,* 261, 4190, 1986.
31. **Stapleton, S. R., Mitchell, D., Salati, L. M., and Goodridge, A. G.,** Triiodothyronine stimulates transcription of fatty acid synthase gene in chick embryo hepatocytes in culture, *J. Biol. Chem.,* 265, 18442, 1990.
32. **Laux, T. and Schweitzer, M.,** Dietary induced pretranslational control of rat fatty acid synthase, *Biochem. J.,* 266, 793, 1990.
33. **Goldman, M. J., Back, D. W., and Goodridge, A. G.,** Nutritional regulation of the synthesis and degradation of malic enzyme mRNA in duck liver, *J. Biol. Chem.,* 260, 4404, 1985.
34. **Fritz, R. S., Stumpo, D. J., and Kletzien, R. F.,** Glucose-6-phosphate dehydrogenase mRNA sequence abundance in primary cultures of rat hepatocytes: effect of insulin and dexamethasone, *Biochem. J.,* 237, 617, 1986.
35. **Kletzien, R. F., Prostko, C. R., Stumpo, D. J., McClung, J. K., and Dreher, K. L.,** Molecular cloning of DNA sequences complementary to rat liver glucose-6-phosphate dehydrogenase mRNA: nutritional regulation of mRNA levels, *J. Biol. Chem.,* 260, 5621, 1985.
36. **Katsurada, A., Iritani, N., Fukuda, H., Noguchi, T., and Tanaka, T.,** Transcriptional and posttranscriptional regulation of malic enzyme synthesis by insulin and triiodothyronine, *Biochim. Biophys. Acta,* 950, 113, 1988.

37. **Katsurada, A., Iritani, N., Fukuda, H., Noguchi, T., and Tanaka, T.,** Effect of aging on induction of rat liver messenger RNA activity for malic enzyme, *Biochem. Biophys. Res. Commun.,* 109, 250, 1982.
38. **Katsurada, A., Iritani, N., Fukuda, H., Noguchi, T., and Tanaka, T.,** Induction of rat liver malic enzyme messenger RNA activity by insulin and by fructose, *Biochem. Biophys. Res. Commun.,* 112, 176, 1983.
39. **Iritani, N., Nagashima, K., Fukuda, H., and Katsurada, A.,** Effects of dietary proteins on lipogenic enzymes in rat liver, *J. Nutr.,* 116, 190, 1986.
40. **Iritani, N., Fukuda, H., and Fukuda, E.,** Age-dependent modification of lipogenic enzymes, *Biochim. Biophys. Acta,* 665, 636, 1981.
41. **Nakanishi, S. and Numa, S.,** Purification of rat liver acetyl coenzyme A carboxylase and immunochemical studies on its synthesis and degradation, *Eur. J. Biochem.,* 16, 161, 1970.
42. **Fukuda, H., Katsurada, A., and Iritani, N.,** Effects of dietary nutrients on substrate and effecter levels on lipogenic enzymes, and lipogenesis from tritiated water in rat liver, *Biochim. Biophys. Acta,* 878, 200, 1986.
43. **Mullar, W. A., Fallona, G. R., Aquilar-Parada, E., and Unger, R. H.,** Abnormal alpha-cell function in diabetes: response to carbohydrate and protein ingestion, *N. Engl. J. Med.,* 283, 109, 1970.
44. **Rocha, D. M., Faloona, G. R., and Unger, R. H.,** Glucagon-stimulating activity of 20 amino acids in dogs, *J. Clin. Invest.,* 51, 2346, 1972.
45. **Steiner, A. L., Ferrendelli, J. A., and Kipnis, D. M.,** Radioimmunoassay for cyclic nucleotides: III. Effects of ischemia changes during development and regional distribution of adenosine 3′, 5′-monophosphate and guanosine 3′,5′-monophosphate in mouse brain, *J. Biol. Chem.,* 247, 1121, 1972.
46. **Rudack, D., Davis, B., and Holten, D.,** Regulation of rat liver glucose-6-phosphate dehydrogenase levels by adenosine 3′,5′-monophosphate, *J. Biol. Chem.,* 246, 7823, 1971.
47. **Watanabe, A., Takesue, A., and Taketa, K.,** Prevention of dietary induction of rat liver glucose-6-phosphate dehydrogenase by cyclic adenosine 3′,5′-monophosphate and its elimination by glucose prefeeding, *Enzyme,* 21, 436, 1975.
48. **Iritani, N., Fukuda, H., and Matsumura, Y.,** Lipogenic enzyme gene expression in rat liver during development after birth, *J. Biochem., (Tokyo),* 113, 519, 1993.
49. **Iritani, N., Fukuda, H., and Matsumura, Y.,** Effects of corn oil- and fish oil-supplemented diets on phospholipid fatty acid composition of rat liver nuclei, *Biochim. Biophys. Acta,* 963, 224, 1988.
50. **Fukuda, H., Tatsumi, H., Iritani, N., Katsurada, A., Kinoshita, C., and Fujita, H.,** Lipogenic enzyme induction and incorporation of exogenous fatty acid into liver and liver nuclei, *Biochim. Biophys. Acta,* 960, 19, 1988.
51. **Limatta, M., Towle, H. C., Clarke, S., and Jump, D. B.,** Dietary polyunsaturated fatty acids interfere with the insulin/glucose activation of L-pyruvate kinase gene expression, *Mol. Endocrinol.,* 8, 1147, 1994.
52. **Jump, D. B., Clarke, S. D., MacDougald, O., and Thelen, A.,** Polyunsaturated fatty acids inhibit S14 gene transcription in rat liver and cultured hepatocytes, *Proc. Natl. Acad. Sci, U.S.A.,* 90, 8454, 1993.
53. **Kimberly, Z. L. and Towle, H. C.,** Carbohydrate regulation of the rat L-pyruvate kinase gene requires two nuclear factors: LF-A1 and member of the c-myc family, *J. Biol. Chem.,* 268, 12787, 1993.
54. **Bergot, M.-O., Diaz-Guerra, M.-J. M., Puzenat, N., Raymondjean, M., and Kahn, A.,** Cis-regulation of the L-pyruvate kinase gene promoter by glucose, insulin and cyclic AMP, *Nucl. Acids Res.,* 20, 1871, 1992.
55. **Shih, H. and Towle, H. C.,** Definition of the carbohydrate response element of the rat S14 gene, *J. Biol. Chem.,* 269, 9380, 1994.

56. **Iritani, N., Hosomi, H., Fukuda, H., and Ikeda, H.,** Polyunsaturated fatty acids regulation of lipogenic enzyme gene expression in liver of genetically obese rat, *Biochim. Biophys. Acta,* 1255, 1, 1995.
57. **Iritani, N. and Fukuda, H.,** Polyunsaturated fatty acid-mediated suppression of insulin-dependent gene expression of lipogenic enzymes in rat liver, *J. Nutr. Sci. Vitaminol.,* 41, 207, 1995.
58. **Moustaid, N., Beyer, R. S., and Sul, H. S.,** Identification of an insulin response element in the fatty acid synthase promoter, *J. Biol. Chem.,* 269, 5629, 1994.
59. **Wolf, S. S., Hofer, G., Beck, K.-H., Roder, K., and Schweizer, M.,** Insulin-responsive regions of the rat fatty acid synthase gene promoter, *Biochem. Biophys. Res. Commun.,* 203, 943, 1994.
60. **Fukuda, H., Katsurada, A., and Iritani, N.,** Nutritional and hormonal regulation of mRNA levels of lipogenic enzymes in primary cultures of rat hepatocytes, *J. Biochem. (Tokyo),* 111, 25, 1992.
61. **Volpe, J. J. and Vagelos, P. R.,** Mechanisms and regulation of biosynthesis of saturated fatty acids, *Physiol. Rev.,* 56, 333, 1976.
62. **Winberry, L. and Holten, D.,** Rat liver glucose-6-phosphate dehydrogenase: dietary regulation of the rate of synthesis. *J. Biol. Chem.,* 252, 7796, 1977.
63. **Bailey, E., Taylor, C. B., and Bartley, W.,** Effects of dietary carbohydrate on hepatic lipogenesis in the rat, *Nature,* 217, 471, 1968.
64. **Fukuda, H. and Iritani, N.,** Effects of high-fructose diet on lipogenic enzymes and their substrate and effector levels in diabetic rats, *J. Nutr. Sci. Vitaminol.,* 29, 691, 1983.
65. **Munnich, A., Lyonnet, S., Chauvet, D., Van Schaftingen, E., and Kahn, A.,** Differential effects of glucose and fructose on liver L-type pyruvate kinase gene expression in vivo, *J. Biol. Chem.,* 62, 17065, 1987.
66. **Cognet, M., Bergot M-O., and Kahn, A.,** Cis-acting DNA elements regulating expression of the liver pyruvate kinase gene in hepatocytes and hepatoma cells, *J. Biol. Chem.,* 266, 8679, 1991.
67. **Noguchi, T., Okabe, M., Wang, Z., Yamada, K., Imai, E., and Tanaka, T.,** An enhancer unit of L-pyruvate kinase gene is responsible for transcriptional by dietary fructose as well as glucose in transgenic mice, *FEBS Lett.,* 318, 269, 1993.
68. **Berghe, G. V. D.,** Metabolic effects of fructose in the liver, *Curr. Top. Cell. Regul.,* 13, 97, 1978.
69. **Sillero, A., Sillero, M. A. G., and Sols, A.,** Regulation of the level of key enzymes of glycolysis and gluconeogenesis in liver, *Eur. J. Biochem.,* 10, 351, 1969.
70. **Takeda, Y., Inoue, H., Honjo, K., Tanioka, H., and Daikuhara, Y.,** Dietary response of various key enzymes related to glucose metabolism in normal and diabetic rat, *Biochim. Biophys. Acta,* 136, 214, 1967.
71. **Gunn, J. M. and Taylor, C. B.,** Relationships between concentration of hepatic intermediary metabolites and induction of the key glycolytic enzymes in vivo, *Biochem. J.,* 136, 455, 1973.
72. **Gavin, L. A., MacMahon, F. A., and Moeller, M.,** The mechanism of impaired T_3 production from T_4 in diabetes, *Diabetes,* 30, 694, 1981.
73. **Noguchi, T., Inoue, T., and Tanaka, T.,** Transcriptional and post-transcriptional regulation of L-pyruvate kinase in diabetic rat liver by insulin and dietary fructose, *J. Biol. Chem.,* 260, 14393, 1985.
74. **Iritani, N., Kukuda, H., Tada, K., Itoh, A., and Noguchi, T.,** Diet differentially regulates glucokinase and L-type pyruvate kinase gene expression in rat liver, *J. Nutr.,* 125, 945, 1995.
75. **Jump, D. B., Clarke, S. D., Thelen, A., and Limatta, M.,** Coordinate regulation of glycolytic and lipogenic gene expression by polyunsaturated fatty acids, *J. Lipid Res.,* 35, 1076, 1994.

Chapter 8

DIABETES MELLITUS: A MITOCHONDRIAL GENOMIC ERROR?

Carolyn D. Berdanier

CONTENTS

INTRODUCTION

Diabetes mellitus is not a single disease, but is a collection of diseases having in common an abnormal glucose–insulin relationship typically expressed as an abnormal glucose tolerance. In normal nondiabetic fasted individuals, administration of a glucose challenge elicits a typical rise then fall in blood glucose levels as shown in Figure 1.

Abnormal glucose tolerance as a diagnostic test for diabetes can vary: there may be a departure from the normal fasting blood glucose level (4-6 mmol/l) and/or post challenge values may be excessively high (exceeding 8 mmol/l) and/or a failure to return to the pre-challenge blood glucose level by 120 min after the challenge. Blood insulin values may exceed normal in some individuals. These individuals are referred to as hyperinsulinemic. They may be hyperinsulinemic yet have normal blood glucose levels and normal glucose

0-8493-9485-6/96/$0.00+$.50

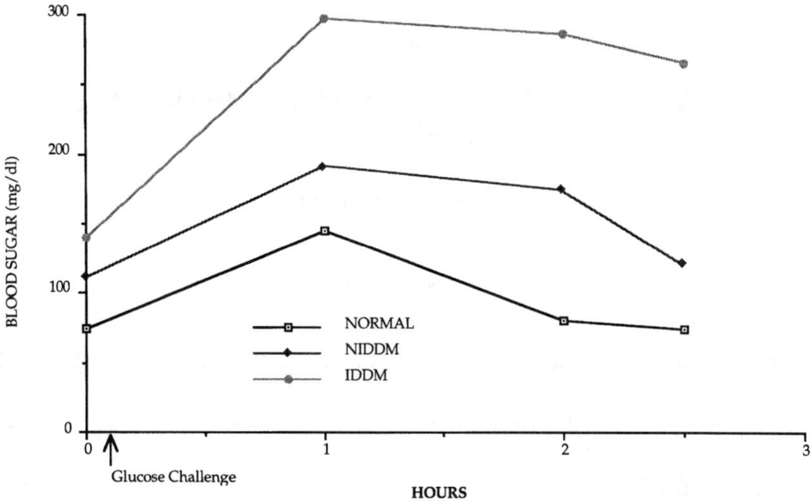

FIGURE 1. Typical glucose tolerance values for normal people and for people with either NIDDM or IDDM. Note that the normal response to a glucose challenge includes a rise followed by a fall in the blood glucose. The NIDDM patient may have a normal fasting blood glucose but the rise following the challenge is greater than the normal and the fall is barely perceptible. The patient with IDDM may have higher values at all time points. Within each group of subjects there may be considerable variation.

tolerance. There are many variants in the response to a glucose challenge. However, because the use of glucose is critical to almost every cell in the body, any deviation in any of the many metabolic steps involved in its use is regarded as suspect with respect to disease development. Because each of these steps involves a protein be it an enzyme, a transporter or a receptor, mutations in the genes for any one of these proteins could occur and explain the impaired glucose use. The candidate gene approach has become quite popular over the last few years as diabetes researchers have mastered the techniques of biotechnology. Listed in Table 1 are some of the candidate genes so far identified or suggested as causes of diabetes mellitus. Before these are discussed, the various forms of diabetes mellitus will be reviewed. This will be followed by an in-depth presentation of diabetes mellitus that can be attributed to defects in the mitochondrial genome.

CONSEQUENCES OF INSULIN DEFICIENCY

The consequences of insulin deficiency have been widely studied in animals exposed to β islet cell cytotoxic agents. These agents destroy the islet cells, and, depending on the dose administered, mild, moderate, or severe deficient states can be induced. The literature on this aspect of diabetes research is voluminous. Two agents, streptozotocin and alloxan, have been used extensively. Streptozotocin is the preferred agent because it is highly specific for the insulin-producing islet cells. It has a very short half life (less than 15 min once

TABLE 1
Candidate Genes in the Pathophysiology of Diabetes Mellitus

Gene/Gene Product	Function	Comment
Glucokinase β cell and liver	Phosphorylates glucose in islet cell. Requires ATP.	Acts as glucose sensor in the islet cell. Several mutations have been found.
MHC	Responds to antigens.	Inappropriate response leads to autoimmune destruction of β cells
Mobile Glucose Transporter	Transports glucose from plasma membrane to cytosol.	Recruited when insulin, bound to receptor site, sends signal to transporter depot; five different transporters have been identified and mutations in the corresponding genes have been reported.
Insulin	Facilitates glucose use, stimulates expression of many genes.	Several mutant forms have been found in addition to mutations in the insulin release and post-translational modification mechanisms.
Insulin receptor	Binds insulin to plasma membrane.	Several mutations have been identified.
mtDNA, $tRNA^{Leu(UUR)}$	Code for leucine in mitochondrial protein.	Lies within the DNA segment responsible for transcription termination of rRNA genes. May also be responsible for the disease, MELAS. Two mutations in this gene have been found.
mtDNA, ATPase 6	Transfers the energy generated by the respiratory chain to F_1ATPase for ATP synthesis	Results in an asparagine substitution for aspartic acid; in turn, a small decrease in ATP synthesis efficiency results. A bp substitution mutation and a duplication mutation have been reported.

in solution) and does not seem to attack other cells in the body that use glucose. In contrast, alloxan, an older drug, not only ablates the islet cells but can also cause kidney and liver damage. When streptozotocin is used in multiple low doses, a mild diabetes can be produced that resembles noninsulin-dependent diabetes mellitus (NIDDM) in many respects.[1-3] Other toxins also can destroy the pancreatic islet cells, but these toxins lack the specificity offered by streptozotocin and will destroy other organs as well.

Insulin deficiency in animals or humans is manifested by body wastage and metabolic abnormalities. The body fat stores are raided, but because insulin is needed to complete fatty acid oxidation, this oxidation is incomplete. As a

TABLE 2
Insulin Effects on Metabolic Pathways

Pathway	Effect	Mechanism
Glycolysis	Increase	Dephosphorylation of glucokinase glucose 2,6 bisphosphatase, pyruvate kinase, pyruvate dehydrogenase, 6-phosphofructose 2 kinase; stimulation of glucose transporter activity
Gluconeogenesis	Decrease	↓ [glucose 2,6 bisphosphate]
cAMP cycle	Alters flow	Increase phosphodiesterase through Ca^{++} mediated phosphorylation; decrease cAMP-dependent potein kinase
Lipogenesis	Increase	↑ synthesis of acetyl CoA carboxylase and fatty acid synthetase ↑ mRNA translation for lipogenic enzymes ↑ phosphorylation of acetyl CoA carboxylase; dephosphorylates HMG CoA reductase and triacylglycerol lipase
Tyrosine kinase	Increase	≠ Vmax of reaction, ≠ [Ca^{++} Mn^{++}] ↓ Km for ATP

result, acetone, β hydroxybutyrate, and acetoacetate, products of incomplete fatty acid oxidation, accumulate. Elevated blood levels of these ketones are observed, as is an elevated urinary excretion. Expired air is also ketone-laden. Rising levels of ketones increase the need for buffering power since ketones are acidic. Not only are the fat stores raided, but so too is the body protein. Proteolysis is enhanced and amino acids thus liberated are used for energy or as substrates for glucose synthesis. The amino groups released as a product of the deamination of these amino acids assist in the buffering of the accumulating ketones. However, ammonia in large quantities is cytotoxic, so the body must increase its capacity to convert it to urea. Insulin-deficient animals (and humans) thus are characterized by loss in body protein, an increase in blood and urine levels of ammonia, an increase in urea synthesis, a negative nitrogen balance, a loss in fat store, elevated blood and urine levels of glucose, and elevated levels of fatty acid oxidation products. Excess thirst and urination also characterize the person with insulin-deficient diabetes. All of these signs and symptoms can be reversed by hormone replacement, hence the term insulin-dependent diabetes or IDDM.

Through the use of insulin-deficient animals the direct and indirect actions of insulin have been elucidated. Listed in Table 2 are some of the many enzymes, carriers, transporters, and metabolic steps that have been shown to be markedly altered in streptozotocin-diabetic rats. Almost every major metabolic process in the body is affected when insulin deficiency occurs. For example, insulin is an important regulator of glycolysis and lipogenesis. Not only because insulin facilitates the uptake of glucose by the cell, but also because insulin, once bound to the plasma membrane receptor, elicits several signals that result in the activation of a large number of anabolic enzymes as well as in the suppression, or down-regulation of, the activities of a large number of catabolic enzymes. Through this signaling mechanism, insulin sets the stage for the enhancement of protein synthesis. It has this effect because it facilitates the provision of energy needed for this energy-expensive process and because it stimulates (indirectly) the translation of mRNA into protein. DNA transcription is also enhanced in the insulin-sufficient cell. Thus, examination of Table 2 reveals changes in enzyme activities or in metabolic processes not only because of the influence of the hormone on glucose use, but also due to its effect on these intracellular signals and on *de novo* mRNA and protein synthesis.

PANCREATIC β ISLET CELL FAILURE

Pancreatic insulin production failure occurs as a result of destruction of the β cells. This destruction can be due to autoimmune disease where the β cell proteins are mistakenly recognized as foreign proteins, or due to viral destruction. In each instance, the diabetic state is due to a failure or an inappropriate function of the immune system. This failure or inappropriate function is an inherited trait. However, because the immune system is a multicomponent one, more than one genetic trait for inappropriate function (autoimmune disease) or failure (inability to combat a viral infection) could account for the disease. Some humans with IDDM have many of the symptoms of autoimmune disease.[4] Autoimmune disease, like diabetes mellitus is a broad category of diseases which includes thyroiditis (inflammation of the thyroid gland), psoriasis, and rheumatoid arthritis, in addition to IDDM. Autoimmune disease arises from mutation(s) in the genes that encode the various components of the immune system. In the instance of IDDM, the genes of the major histocompatibility complex (MHC) are suspect. More than 100 genes are involved in an area of chromosome 6 that is approximately 2×10^6 nucleotides in length.[5-8] In humans with autoimmune IDDM, antibodies to islet cell cytoplasmic proteins have been reported.[9] Glutamic acid decarboxylase (GAD), an enzyme found in β cells that catalyzes the synthesis of the neurotransmitter γ aminobutyric acid (GABA), is just one of these self-antigens that elicits an auto antibody response. Antibodies to other self proteins have also been identified and in many instances these antibodies appear well before the manifestation of the clinical state. As well, the BB rat and the NOD mouse, animals that develop severe IDDM, also manifest symptoms of a utoimmune disease[10-18] and these symptoms are repressed by the immunosuppressive drug cyclosporin. This immunosuppression results in a lessening of subsequent disease severity. Studies of the inheritance of the diabetic tendency and the inheritance of human leukocyte antigens (HLA) show that these A, B, and C genes encode antigens and that aberrations in these segregate in the same pattern as IDDM. HLA participate in T-cell recognition and in the rejection of a variety of aberrant cell proteins such as proteins from malignant cells or viral infected cells. It can also include certain self-cell types. Of interest is the fact that some years ago, scientists at the Jackson Laboratory, using traditional gene mapping techniques to identify the inheritance of the diabetic trait in the db/db mouse, reported that this trait was carried on chromosome 4. This is not the same chromosome that carries the codes for antigen recognition.

DIABETES SECONDARY TO VIRAL INFECTIONS

Studies of IDDM in human and rodent populations show that IDDM can be preceded by communicable viral diseases.[19-32] IDDM in rodents has been induced by exposure to two closely related viruses, the encephalomyocarditis virus and the Coxsackie B virus.[20-25,27,29-32] Not all rodents or humans respond to these viruses by developing IDDM. Some individuals are responsive while others are not. This suggests a genetic determination of susceptibility to these infections that probably involves genes that code for the various components of the immune system.[5] Notkins and co-workers[24-28] have demonstrated that susceptibility is compatible with a single gene acting in an autosomal recessive manner. The mode of action of such a gene would be to control specific virus receptors on the β cell membrane or perhaps to control the membrane permeability to such viruses. The gene would thus facilitate the incorporation of the viral DNA into the β cell DNA and destroy it. Fluorescein labeling of viral antibody has confirmed the entry of the virus into the β cell[5,27] and Coxsackie B4 virus-induced changes in β cell mRNA have been documented.[23] Evidence of virus-induced IDDM in man has been gathered through post-mortem studies of pancreatic tissue excised from children with fatal viral infections caused by a variety of viruses. Of the 250 children studied, only 7 had a Coxsackie B infection, and of these, 4 had significant β cell destruction and evidence of acute or chronic inflammation in the islet tissue.[32]

NONINSULIN-DEPENDENT DIABETES MELLITUS

In contrast to IDDM, the incidence and prevalence of NIDDM fluctuates broadly. Food abundance and economic status of the population are factors which influence NIDDM development. During the worldwide depression in the 1930s, many people were without sufficient income to purchase an abundance of energy-rich food. As a result, their food selection was restricted to the lower priced food items and they restricted their total food intake to conserve their limited financial resources. This food restriction resulted in decreased body size and fatness. In turn, a reduction in body fatness as well as food intake was associated with a reduction in the number of new cases of NIDDM.

The range of disorders in man and laboratory animals that fit the description for NIDDM is far greater than that for IDDM. For most of the NIDDM disorders, obesity is a concurrent feature. This obesity either precedes or accompanies the development of abnormal glucose tolerance.

The genetic influence on the development of diabetes mellitus is stronger for NIDDM than for IDDM. In studies of identical twins for example, concordance for IDDM was 60% whereas for NIDDM, the concordance was 100%. NIDDM, however, is more responsive to environmental influences than is IDDM, and it has been said that it is a disease that develops as a result of an interaction between the genetic heritage of the individual and that individual's lifestyle choices. The age of onset as well as the severity of NIDDM is thus the product of this interaction. Whereas IDDM management includes daily hormone replacement therapy, many patients with NIDDM can manage their disease with changes in lifestyle: decreased food intake, increased physical activity, reduced body fat stores, and stress avoidance. Most patients with NIDDM can manage to avoid the need for daily insulin injections if they keep their weight down and their activity levels up. However, as they age, or if they become obese or seriously ill, such hormone replacement therapy may become necessary. Whereas the reasons for IDDM development are confined to the islet β cell and its failure, NIDDM can develop in response to variety of abnormalities in the structure of insulin, its transport to the target tissue, target tissue receptor abnormalities, and to abnormalities in one or more of the metabolic pathways needed to effectively use the glucose molecule. We are only now beginning to appreciate the tremendous variability in the pathophysiology of NIDDM. In part, this has been due to the many detailed studies of spontaneously diabetic animals.[33] A number of mutations have been recognized in both large and small laboratory animals, and identification of these animals has made possible the identification of similar problems in man.

Several mutations in man have been identified that are associated with NIDDM. Beginning with the pancreas, it has been found that the islet cell may not be producing insulin in the appropriate amino acid sequence to have full biological activity. There are a number of cis- and trans-acting factors that convey β cell specificity to the insulin gene.[34] Should one or more of these be absent or nonfunctional, the β cell will not transcribe the insulin gene. This is not a frequent occurrence. Insulin gene mutations have been documented in both man and mouse that result in a variety of amino acid substitutions that have a variety of effects on insulin action.[35,36] Steiner, et al., for example, studied a number of families with aberrant insulin genes.[35] Ten families were studied that had single point mutations that resulted in amino acid substitutions in the proinsulin molecule. Six of these substitutions resulted in the secretion of defective insulin molecules due to changes within the A or B chains. Replacing A3-Val with Leu, B24-Phe with Ser, or B25-Phe with Leu resulted in molecules that were immunoreactive but did not bind to the plasma membrane insulin receptor site on the adipocyte. Four additional families were found to have insulin gene mutations that prevented the recognition of the C peptide A chain dibasic cleavage site, by the appropriate β cell processing protease. This mutation is characterized by high levels of circulating proinsulin. Such a mutation has been studied in transgenic mice by the introduction of this mutant gene into the germ line, resulting in its expression of high levels of proinsulin along with the normal mouse insulin gene expression by the islet cell.[35] The variability of the insulin gene in a group of phenotypically normal and diabetic individuals has been studied.[36] Although several variants have been found, none were considered abnormal. That is, the amino acid substitutions were in places that did not affect the conformation of the molecule or its active site, or the cleavage of proinsulin to active insulin. The variety of base pair substitutions and subsequent amino acid substitutions in the insulin molecule appears to be large, yet the impact of these aberrations (if they can be called aberrations) is very small. However, should substitutions occur that affect insulin conformation or proinsulin cleavage, or reactive sites in the insulin molecule, then diabetes would develop. Whether the diabetes is insulin-dependent or noninsulin-dependent, would as mentioned, depend on the nature of the defect. Hyperproinsulinemia has been found in obese mice that have a mutation (on chromosome 8) in the gene for carboxypeptidase E. This is the enzyme needed for C peptide cleavage.[37] This defect is separate from the above in that the defect is not in the insulin gene per se but in a gene that affects post-translational processing.

Recently there has developed an hypothesis that islet cell failure is due to a mutation in the gene for glucokinase.[38-42] Aberrant glucokinase gene expression would mean that glucose as a potent stimulator for insulin release would not have this function since its secretogogue action is dependent on its phosphorylation via glucokinase.[42,43] Several mutations in the glucokinase gene in the β cell have been found.[38] One has been mapped to chromosome 7 and has been linked to IDDM.[39] Another has been found in islet cells in association with NIDDM,[40] while a third, found in hepatic cells but not β cells, has been linked with both NIDDM and obesity.[41] Other mutations were shown to be tightly linked to the development of NIDDM in a group of French families.[42] There is specific tissue regulation of in the glucokinase gene that, if aberrant, results in a decrease in glucose sensing by the β cell. Some of these mutations result in NIDDM rather than IDDM. In addition, certain of the amino acids can act as insulin secretogogues and the action of these nutrients is independent of glucokinase. Defects in the secretion mechanism in addition to impaired glucokinase activity have also been suggested.[44-45] These involve the calcium ion and the phosphatidylinositol second messenger system. However, studies of insulin release mechanism using a cDNA clone encoding a K+ATP channel-like protein showed no association of the different polymorphism with NIDDM.[45]

Among the more common characteristics of NIDDM is the resistance of the peripheral fat cell to the effects of insulin. The plasma membrane receptor on the fat cell may be aberrant in structure or number. If abnormal in structure, insulin may not be bound to it. This may occur either because the receptor protein is abnormal or because the fat cell is enlarged by its stored fat and the receptor protein is distorted. When the fat cell shrinks, i.e., when the obese person or obese animal loses a large percentage of excess stored fat, insulin binding is normalized. Insulin binding, in this instance, is directly related to adipocyte size, which in turn is directly related to the size of its fat store. The insulin resistance, thus observed, refers to the particular abnormality of the fat cell insulin receptor conferred upon it by its excess fat store.[46]

Insulin resistance can, however, occur for other reasons, and it can occur at other tissue sites as well. By definition, insulin resistance means that it takes abnormally large amounts of insulin to regulate or control blood glucose within normal limits. An animal (or human) with insulin resistance may not develop abnormal glucose tolerance *if* the pancreas is able to sustain an abnormally high insulin output. In this circumstance, the individual may be insulin-resistant and hyperinsulinemic but not hyperglycemic or have abnormal glucose tolerance. Eventually, the islet β cells may not be able to sustain this high insulin output and the insulin resistance will then progress to abnormal glucose tolerance and subsequently to the diabetic hyperglycemic state. Insulin resistance may be a feature of the fat cell, and/or the muscle cell, and/or the liver cell and, in each instance, genetic errors in the DNA which code for this particular plasma membrane protein have been found.[47-50] Insulin receptor cDNAs have been cloned.[47-48] Ebina[48] reported that the receptor contained 12 more amino acids in its α subunit than found by Ullrich.[49] Nonetheless, it is generally agreed that the receptor is synthesized as a single polypeptide precursor of 1382 amino acids which contains a signal peptide of 27 amino acids, an α subunit of 735 amino acids (including four basic amino acids of the processing site) and a β subunit of 620 amino acids. After glycosylation, processing, and disulfide bonding, it is expressed as a heterotetramer composed of two α subunits with a molecular weight of 95,000. The α subunit is entirely located on the exterior aspect of the plasma membrane while the β subunit extends through the plasma membrane into the cytoplasm. The human insulin receptor gene is located on chromosome 19. It consists of 22 exons which span 120 kb. The α subunit is encoded by the first 120 kb and includes a signal peptide, an insulin binding region, and a cysteine rich region. The β subunit is encoded by the last 30 kb. As mentioned, this subunit is the portion of the receptor molecule that crosses the plasma membrane. It has a proreceptor processing site and the final portion of the unit of the receptor molecule consists of tyrosine kinase. When insulin binds to the α subunit, autophosphorylation occurs, resulting in activation of tyrosine kinase. If the DNA coding for the receptor protein has mutated such that lysine, an important amino acid for ATP binding and receptor activity, is replaced by arginine, alanine, or methionine, the receptor is no longer able to mediate insulin action. Thus, lysine is a critical component of the 1030 ATP binding site. Clusters of tyrosine residues are also critical to the autophosphorylation process. If tyrosine is replaced by phenylalanine, again receptor activity is compromised. Other mutations in the code for the receptor protein which affect its amino acid sequence have been reported.[47-50] These mutations (depending on their location) likewise can affect the activity of the receptor and, in turn, can explain insulin resistance.

Insulin resistance can also be attributed to post-receptor defects. Errors in the codes for the mobile glucose transporter (called GLUT 1,2,3,4, and 5) have received considerable attention,[51-56] especially as these errors may relate to the development of obesity. Peripheral glucose uptake may be impaired if the mobile glucose transporter (GLUT 4) in the adipocyte is aberrant.[52] An aberrant adipocyte transporter would result in less insulin-stimulated glucose uptake by these cells, with the result of increased levels of circulating glucose, which in turn would stimulate pancreatic insulin release. Thus, clinically, one could observe both hyperglycemia and hyperinsulinemia. Note the similarity of clinical features between this genetic error and the error described above that involves the gene for the insulin molecule. In both instances, one observes peripheral fat cell resistance together with compensatory increases in hepatic glycolysis, pentose shunt, glycogenesis, lipogenesis, hepatic lipid output, and adipocyte lipid uptake and storage. The compensatory increases in hepatic metabolism are probably driven by both the hyperinsulinemia, with its effects on anabolic processes, and the hyperglycemia, which provides a continuous high level of substrate for these anabolic pathways. Studies of obese diabetic Zucker (ZDF/drt) rats and obese diabetic A^{vy}/a mice[57] have suggested that the substrate glucose may have regulatory properties with respect to the expression of the gene for the mobile glucose transporters in liver, muscle, and fat cells. This suggestion was based on the observation that alterations in GLUT 2 and/or GLUT 4 proteins were not associated with the obesity per se, but were secondary to severe hyperglycemia that was the result of the diabetic state. Incidentally, studies of GLUT 1 and GLUT 4 in adipocytes from normal rats[58] showed that feeding a high glucose diet increases GLUT 4 activity while feeding a high fat diet isocalorically, decreased GLUT 4 activity and increased GLUT 1 activity. These results indicated that the two transporters differed in their response to insulin and glucose. In normal nondiabetic rats, GLUT 1 activity is probably responsible for basal glucose uptake while GLUT 4 is under the control of insulin. When insulin levels are high, GLUT 4 activity is high, whereas GLUT 1 seems to be independent of the hormone, but responsive to the nutritional state of the body. It also appears from studies of normal and genetically obese and/or diabetic animals that the multiplicity of glucose transporters, while seemingly redundant, comprises a "fail-safe" mechanism to ensure glucose disposal. Should one of the isoforms be aberrant because of a genetic error, there would be a backup transporter that could increase in activity to compensate for the genetically determined loss.

The study of dietary factors which affected GLUT 1 and GLUT 4 activity also provides some insight into the nature of nutrient–gene interactions that may explain NIDDM. As mentioned, lifestyle choices can affect the time course and severity of NIDDM. Lifestyle choices include decisions about the quantity and composition of the food consumed as well as the level of physical activity. As indicated above, the diet composition can affect the recruitment of GLUT 1 and GLUT 4 transporters. If the NIDDM patient had an error in one or the other of these transporters, clearly, the choice of the diet would be critical in the disposal of glucose by the adipocyte and in turn critical to the regulation of glucose homeostasis.[59] In addition, exercise has been shown to affect the regulation of glucose transport in skeletal muscle.[59] With exercise, the number of GLUT 4 transporters increases, independent of insulin status. This explains the decrease in insulin need by diabetic athletes, an observation frequently made but poorly understood in practical terms. With exercise, the muscle uses glucose without the need for insulin bound to the receptor. Insulin is thus not needed for the recruitment and translocation of GLUT 4 transporters. With this in mind, if NIDDM develops as a result of an error in the insulin receptor, then exercise will promote glucose use and facilitate the reduction in blood glucose. If the person thus includes daily exercise and avoids high-energy and/or high carbohydrate diets, then the phenotypic expression of the genotype for either a receptor or transporter error might be postponed or avoided. In turn, one can understand why the incidence of NIDDM decreases with restricted food supplies and increases in physical activity, as happens during times of economic duress or war.

OXIDATIVE PHOSPHORYLATION AND DIABETES MELLITUS

Both the oxidation of glucose and the release of insulin are energy-dependent processes requiring ATP. The phosphorylation of glucose in the liver and the islet cell is catalyzed by the ATP-independent enzyme, glucokinase. This enzymatic step is presumed to be the glucose sensor of the islet cell.[41,42] However, if the amount of ATP is limited in the islet cell due to decreased synthesis via oxidative phosphorylation (OXPHOS) this would have the effect of impaired responsiveness of the pancreas to a glucose challenge. Furthermore, protein synthesis is energetically very expensive. Insulin is constantly being synthesized by the islet cell. Should there be an impairment in ATP synthesis, insulin synthesis will likewise be impaired. Thus, a genetic error in any one of the many proteins which comprise OXPHOS could result in an impaired synthesis of ATP.

OXPHOS occurs in the mitochondrion. It consists of the respiratory chain and the ATP synthase (F_1F_0ATPase). Respiration is the process by which most cells obtain energy from the oxidation of fats and carbohydrates. Carbon dioxide and water plus the energy trapped in the ATP molecule or released as heat are the products of this process.

Glycolysis as well as fatty acid and amino acid oxidation result in the production of an activated two-carbon residue, acetyl CoA. This common metabolic intermediate is joined to oxalacetate to form citrate and is then processed by the citric acid cycle. This cyclic sequence of reactions is found in every cell type that possesses mitochondria. The cycle is catalyzed by a series of enzymes and yields reducing equivalents (H^+) and carbon dioxide. Reducing equivalents are produced when α ketoglutarate is produced from isocitrate, when α ketoglutarate is decarboxylated to produce succinyl CoA, when succinate is converted to fumarate, and when malate is converted to oxalacetate. These reducing equivalents, one pair for each step, are carried to

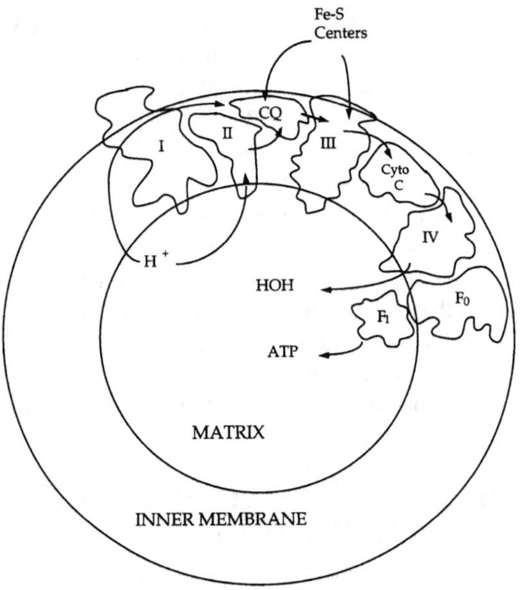

FIGURE 2. Schematic representation of oxidative phosphorylation where Complex I consists of all the NAD-linked reactions which donate electrons to the first series of iron sulfur centers of the chain. Complex II consists of all the FAD-linked reactions which donate electrons via ubiquinone (Q). Complex III consists of cytochrome b562, b566, iron sulfur to cytochrome C1 electron transfers. Complex IV consists of the cytochrome a and a3 electron transfers. Complex I has 39 subunits; Complex II has 4 subunits; Complex II has 10 subunits, and Complex IV has 13 subunits. Complex V is the ATP synthase/ATPase complex and consists of an hydrophobic segment (F_0) and an hydrophillic segment (F_1). The F_1 consists of 5 nonidentical protein subunits and is water soluble. The F_0 consists of 8 subunits and is hydrophobic. F6 is one of these subunits whose activity depends on the fluidity of the phospholipid environment.

the respiratory chain by way of either NAD or FAD. Reducing equivalents from succinate are carried by FAD while those produced by the oxidation of the other substrates are carried by NAD. Once oxalacetate is formed, it can then pick up another acetate group from acetyl CoA and begin the cycle once again by forming citrate. Thus, for every turn of the cycle, two carbon dioxide molecules and eight pairs of reducing equivalents are produced. As long as there is sufficient oxalacetate to pick up the incoming acetate, the cycle will continue to turn and reducing equivalents will continue to be produced and these, in turn, will be joined with molecular oxygen to produce water, the end product (with carbon dioxide) of the catabolic process.

RESPIRATORY CHAIN

The pairs of electrons produced at the four steps described above in the citric acid cycle as well as electrons transferred into the mitochondria via other processes are passed down the respiratory chain to the ultimate acceptor, molecular oxygen and in the process, ATP is synthesized. This is called oxidative phosphorylation. The respiratory chain is shown in Figure 2.

The enzymes of the respiratory chain are embedded in the mitochondrial inner membrane. They catalyze the series of oxidation–reduction reactions which comprise the respiratory chain. As electrons pass down the respiratory chain, ATP is synthesized. The mechanism that links the respiratory chain to ATP synthesis has been discussed at length. Perhaps the best accepted mechanism is that proposed by Mitchell in 1968.[60] Mitchell proposed that the sequential reactions of the respiratory chain generate an electrochemical gradient of H^+ ions across the inner mitochondrial membrane. According to Mitchell, this gradient serves as a means for coupling the energy flow from electron transport to the formation of ATP. In this proposal, the inner mitochondrial membrane is an integral part of the mechanism which couples the energy gradient of the electron flow in the respiratory chain to the synthesis of ATP. *A priori,* the membrane must be intact and in the form of a continuous closed vesicle for oxidative phosphorylation to take place. If the membrane is disrupted, coupling will not occur. Respiration may occur, but ATP synthesis will not. Mitchell proposed that the enzymes of the respiratory chain are arranged so as to transport hydrogen ions from the matrix across the inner membrane such that the gradients will develop in close proximity to the F_1F_0ATPase complex and provide a proton gradient sufficient to drive ATP synthesis by causing a dehydration of ADP and Pi.

Functionally, the various enzymes of oxidative phosphorylation can be divided into five complexes. Complexes I, II, III, and IV plus ubiquinone (Q) and cytochrome c make up the respiratory chain. The electron carriers are the quinoid structures (FMN, FAD, Q) and the transition metal complexes. At three stages in the chain, oxidative energy is conserved (Figure 2) via coupled vectorial proton translocations and the creation of a membrane proton gradient ($\Delta\mu H^+$). This protonic energy is what drives the synthesis of ATP. Should this energy be dissipated as happens when agents such as dinitrophenol are present, ATP synthesis does not occur or occurs to a very limited extent. Uncoupling occurs because protons are transported through the inner membrane from the intramembrane space to the matrix. This "short circuits" the normal flow of protons through complex 5. In such a circumstance, electrons continue down the respiratory chain but the protonic energy thus generated is not captured in the high-energy phosphate bond. Instead, this energy is released as heat. There are variations or degrees of protonic energy capture *in vivo* which are related to the presence or graded absence of naturally occurring agents, which serve to either inhibit the action of complex V[52] or uncouple the respiratory chain energy generation process from the energy capture process of ATP synthesis. Brown fat cell mitochondria have the ability to synthesize a protein which functions as an uncoupler of oxidative phosphorylation and in so doing generate heat. Brown fat thermogenesis has been suggested by numerous investigators to play a role in the regulation of body fat stores.[58-63] It is reasonable to assume that other cell types such as hepatocytes or muscle cells may also have mechanisms that control the tightness or looseness of coupling. Naturally occurring substances such as long chain fatty acids have some role in this respect. Otto and Ontko[64] for example, showed that the influx of long chain fatty acids into isolated mitochondria resulted in a mild swelling of the organelle and a reduction in the amount of ATP produced. Williams, et al.,[69] as well as others,[70-76] reported a change (decrease) in the ratio of ATP produced to oxygen consumed in rats fed hydrogenated coconut oil or fat-free diets. These reports indicated that the fatty acids in the mitochondrial membrane (which reflected somewhat the fatty acids of the diet) changed the degree to which the protonic energy developed by the respiratory chain could be captured in the synthesis of ATP. That which was not so captured was released as heat. Hence, animals fed diets devoid of essential fatty acids evidence a reduction in fuel efficiency (the gain in weight related to the amount of food consumed) and, although not reported, these rats should also evidence an increase in heat production.

The membranes of the mitochondrion contain phospholipids (PL) that are specific to the particular membrane and may affect the activity of the various proteins embedded in it. The proteins most likely affected are those that depend on the fluidity of PL and their charge. Fluidity is determined by the chain length and saturation of the fatty acids attached at carbons 1 and 2 of the PL, whereas charge is determined by the structure of the phosphorylated group attached to carbon 3. Within the PL bilayer and extending beyond it into the matrix are the protein components that are organized into the respiratory chain complexes, the F_1F_0ATPase, receptors, transporters, transcriptional machinery, structural proteins, and the enzymes of the Krebs cycle and fatty acid oxidation.

FUNCTIONAL CHARACTERISTICS OF MITOCHONDRIA

An isolated preparation of mitochondria is functionally tested by an analysis of its respiration and the coupling of this process to ATP synthesis. Well-coupled mitochondria utilize the chemical energy inherent in the site-specific substrates and produce water and energy which is either released as heat or trapped in the high-energy bounds of ATP. Through the electron transport complexes described above, reducing equivalents are joined to molecular oxygen to make water. In the process, energy is released and ultimately transferred to Complex V, the F_1F_0ATPase. A criteria useful in evaluating coupling is the respiratory control ratio (RC).[77] This is the ratio of the O_2 consumed by isolated mitochondria upon addition of ADP, to that O_2 consumed once all the added ADP is phosphorylated to ATP. Those mitochondria that fail to return to the resting state after being stimulated by the addition of ADP are said to be "uncoupled." If the mitochondria cannot be stimulated to respire, there will be no response to the addition of ADP, that is, there will be no difference between the resting state (state 4) and the stimulated state (state 3). These mitochondria are no longer functionally "alive." Uncoupling can be induced through the use of agents such as dinitrophenol. Inappropriate osmolarity of the media in which the mitochondria are studied can also result in uncoupling. In the living, respiring system there can be variation in the degree of coupling of respiration to ATP synthesis. Indeed, survival is dependent on this variation. Exposure to cold for example, requires heat generation for survival. Without the ability to generate additional heat through selective uncoupling, the animal becomes hypothermic and dies. Many of the genetically obese rodents that may also express NIDDM are unable to respond to cold.[63]

MITOCHONDRIAL GENOME

While most of the proteins involved in OXPHOS are encoded by nuclear genes, 13 are encoded by the mitochondrial DNA (mtDNA). This genome in the human is ~16.6 kb double stranded circle and contains 37 genes. Two are ribosomal RNA, 22 are tRNAs and 13 encode the mRNAs for the OXPHOS proteins. These proteins include seven subunits of Complex I (NADH dehydrogenase-ubiquinone oxidoreductase (ND1, ND2, ND3, ND4, ND5, ND6); one subunit of Complex III (ubiquinone-cytochrome c oxidoreductase, cyto b); three subunits of Complex IV (cytochrome c oxidase, COI, COII, COIII) and two subunits of Complex V (ATP synthase, ATPase 6, ATPase 8). Chapter 1 describes some of the disorders associated with mutations in the mitochondrial genome. Of interest in this chapter are mutations in this genome that phenotypically are expressed as diabetes mellitus.

Several reports[77-80] have recently appeared that associated diabetes mellitus with a mutation in the tRNA for leucine at position 3243 in the mitochondrial genome. Deafness is also associated with this mutation. This mutation is an A→G transition. Besides altering the tRNA sequence, it alters the transcription termination of the ribosomal RNA genes. Transcription termination at this specific site is important in regulating the appropriate steady-state ratios of mRNA to rRNAs. This A→G transition has also been reported in people with MELAS[78] (mitochondrial encephalomyopathy, lactic acidosis, and stroke-like episodes). MELAS is a devastating condition and patients have poor survival. In MELAS the regulation of transcription termination is only 1/10th as effective as that which occurs in normal individuals. Although MELAS is characterized by a shortened lifespan, those that survive to adulthood have had children. The males do not transmit their malady to their children while the females do. There appears to be a threshold level of mutated mtDNA below which the symptoms of MELAS are not obvious and a further threshold level of mutated mtDNA for diabetes mellitus and deafness. This variability of symptoms as well as disease syndrome is indicative of heteroplasmy. Individuals having both normal and mutated $tRNA^{Leu}$ genes are heteroplasmic and the percent mutation determines what symptoms are observed. From mildly impaired glucose tolerance to overt NIDDM and deafness to muscle and nerve abnormalities that characterize the MELAS syndrome, all may be related simply to the relative proportions of normal to mutated $tRNA^{Leu}$. Leucine is an essential amino acid and as such would be a required ingredient for all newly synthesized mitochondrial protein. If the tRNA for leucine was aberrant, one could anticipate a reduction in mitochondrial protein synthesis. Indeed, Moraes et al.[78] have reported that this occurs in MELAS. Similar studies have not been conducted in NIDDM patients with this defect; however, in the report by Kishomoto et al.,[77] it was noted that their patients were leaner than normal with less lean body mass. The Japanese patients also had defective insulin release and required insulin supplementation to control their blood glucose levels within normal limits. Heteroplasmy in various tissues from these patients was observed. The percent mutated DNA was greatest in muscle and least in leukocytes. These observations are consistent with those of van den Ouweland.[80] The Dutch and French populations studied by van den Ouweland et al.[80] had additional mutations in their mtDNA but none were considered consequential. Six of these were likely errors in the Cambridge sequence while the remaining 12 had no effect on the amino acid sequence of the resultant gene products. Six additional bp substitutions were found in the structural and protein coding regions, but none of these segregated with NIDDM upon pedigree analysis. Only the $tRNA^{Leu(UUR)}$ segregated with diabetes in these families. Two transitions in the ATPase 6 gene were found: 8860 A→G which results in a T→A amino acid substitution and at position 8994, and a G→A that translates to L to L amino acid sequence change. Van den Ouweland et al. suggest that these transitions are not relevant to NIDDM development.

Nakagawa et al.[83] have reported another mutation in the $tRNA^{Leu}$ gene that associates with NIDDM. In this instance a G to A mutation at position 3316 was reported. This mutation was more frequent (3.4%) in the Japanese than the 3243 mutation (0.3%).

Ballinger et al.[84] reported a 10.4-kb deletion in mtDNA that associated with insulin-dependent diabetes mellitus and deafness. Later, these investigators reported that this deletion was a duplication in base sequence rather than a deletion of this size. Others[85] have also reported on mtDNA duplications that resulted in diabetes and renal disease as well as cerebellar ataxia. In this instance of duplication, the ATPase 6 gene was affected as was the cytochrome b gene. This resulted in early death for the affected children. Nonetheless, the mutation affected mtDNA replication and protein synthesis was inhibited. In turn, this had striking effects on pancreatic islet function through effects on OXPHOS. As mentioned earlier in this chapter, mutations which result in decreased ATP synthesis will affect both insulin protein synthesis and insulin release. Both are ATP-dependent and aberrations in either or both will result in diabetes mellitus.

Studies of the BHE/cdb rat have shown that rats of this strain mimic the human with NIDDM.[86] Numerous studies of their metabolic pathways as well as characteristic features of the liver, pancreas, and kidney have been conducted. Notable is a fatty liver with a greatly increased lipogenic and gluconeogenic activity.[87,88] The endocrine pancreas is characterized by an age-related decrease in responsiveness to a glucose challenge, deterioration of the β cells, and atrophy.[89] The kidneys show evidence of glomerulosclerosis.[90-91] The average lifespan is 50% less than that of a normal rat and the chief cause of death is renal disease. Along with these metabolic features is a decrease in efficiency of ATP synthesis.[76,92-96]

All of these observations suggested that a mutation might have occurred in genes which encode the components of oxidative phosphorylation. Since increased respiration rates but not ATP synthesis through thyroxine treatment[97] could be obtained, we knew that the components of the respiratory chain were probably normal. What was not obtained with hyperthyroidism was an increase in ATP synthesis. Thus, the hypothesis was that a mutation had to have occurred in the F_1F_0ATPase.

NIDDM-prone BHE/cdb rats have a point mutation in the mtDNA gene that encodes the ATPase subunit 6 of the F_0 portion of the F_1F_0ATPase.[98] The F_1F_0ATPase catalyzes the synthesis of ATP using the proton gradient generated by the respiratory chain and capturing it in the high-energy bond of the ATP. This is the first time such a mutation has been found in a proven NIDDM-prone rat. An identical mtDNA sequence was published[99] for female Sprague-Dawley rats from WIGA, Sulzfeld. This colony in Germany was obtained from Charles River who in turn obtained their rats from Sprague-Dawley, Madison, Wisconsin. The German colony was maintained for several decades without repopulation from either Charles River or Sprague-Dawley. As these rats age they become hyperglycemic, hypertriglyceridemic, and hypercholesterolemic.[100] Their glucose tolerance has not been intensively studied but because of their age-related hyperglycemia,[100] one could assume that it becomes impaired as they age. These characteristics are very similar to what we have reported for the diabetes prone BHE/cdb rat. Since both strains have the same base sequence in the region of the F_0ATPase, we suggest that in both the same spontaneous mutation has occurred. We also suggest that this single mutation is so mild that it has little impact on overall glucose metabolism in the young (50-day-old) rat. In the young rat there likely are a number of compensatory mechanisms operative that serve to keep the blood glucose within normal range. However, as the animals age, decompensation occurs and the regulation of glucose homeostasis becomes less controlled with subsequent measurable aberrations in specific tissues. While the rats from WIGA Sulzfeld have not been followed in the same detail as the BHE/cdb rats, no doubt there are similarities in their age-related changes in glucose metabolism. The mutation in the BHE/cdb rat was found in both young and mature rats. Sprague-Dawley rats were examined at the same ages at the same time. We found four single base pair differences when comparing the sequences of the BHE/cdb rats to those of the Sprague-Dawley rats. The base sequences in the Sprague-Dawley rats were identical to those of Gadalata et al.[101] Those of the BHE/cdb rats were not, but were identical to those reported for the WIGA Sprague-Dawley rats.[99] Three of the nucleotide substitutions (at bp 402, 571, 876) were silent, and should have no effect on the amino acid sequence of the resultant protein. The nucleotide substitution at bp 523 probably is not silent. At this position the BHE/cdb mtDNA contained an adenine while the Sprague-Dawley mtDNA contained a guanine. The Sprague-Dawley sequence (GAC) provides the codon for aspartic acid while the BHE/cdb sequence (AAC) encodes asparagine. This means that there probably is an amino acid substitution in the F_0ATPase. The substitution we infer from this base substitution in the mtDNA would place an asparagine in a critical portion of the F_0 molecule. This is the portion of the molecule that forms a polar pocket through which the protons flow. This pocket spans the membrane and is surrounded by the membrane lipid. Having a side chain with both a keto group and an amino group (asparagine) instead of an acidic carboxyl group (aspartate) at this spot would result in a reduction in efficiency, in the synthesis of ATP, and would provide the basis for a slip or leak in the proton gradient that is generated by the respiratory chain.[102-106]

The mitochondria have a dual bilayer membrane structure that surrounds the oxidative phosphorylation system. The lipids in these membranes serve to regulate the activity of the membrane-bound components. This is probably of importance to the F_0 which must have some degree of flexibility within its lipid environment for maximal function. Thus, the fluidity of the lipid bilayer is a modifying factor in the synthesis of ATP using the energy of the proton gradient developed by the respiratory chain. In our studies of BHE/cdb rats fed diets containing different kinds and amounts of fat, we noted that feeding hydrogenated coconut oil in a sucrose-rich diet resulted in far less efficient coupling of respiration to ATP synthesis and indeed, what might be regarded as a minimal strain difference in coupling in rats fed a control diet became a major difference when the composition of the diet was manipulated. Normal animals fed these diets did not exhibit the drastic changes in mitochondrial activity observed in the BHE/cdb rats. Furthermore, when these rats were fed the coconut oil diet and were treated with thyroxine, which should have increased coupling efficiency, the reverse occurred.[97] More mitochondrial protein was synthesized and this would include the F_0ATPase, but because the F_0ATPase had an abnormal amino acid sequence it could not effectively capture the energy released by the increased respiratory chain activity.

The studies in rats showed that diet composition affected OXPHOS and that this could be interpreted as a nutrient–gene interaction effect vis é vis a mutation that ultimately manifests itself as NIDDM. This may also apply to humans. As outlined in the introduction to this chapter, the incidence of new cases of NIDDM declines during periods of restricted food intake. If the human is polymorphic with respect to the mtDNA ATPase gene, and likely this is the case, then NIDDM will only develop when food intake is unrestricted and is especially rich in sugars and fats. This is an instance where a mutation is not pathologic unless or until the diet is such that the metabolic compensations ultimately prove difficult.

SUMMARY

Diabetes mellitus is a disorder of inappropriate regulation of glucose metabolism. The symptoms are various and thus the disorder is not a single one, but rather is a collection of disorders. A number of candidate genes have been proposed because the genetic linkages of all of the disorders called diabetes mellitus are very strong. Both nuclear and mitochondrial genes have been identified.

REFERENCES

1. **Wolf, J., Lilly, F., and Shin, I.,** The influences of genetic background on the susceptibility of inbred mice to streptozotocin induced diabetes, *Diabetes,* 33, 567, 1984.
2. **Evoy, R. C., Andersson, J., Sandler, S., and Hellerstrom, C.,** Multiple low dose streptozotocin-induced diabetes in the mouse, *J. Clin. Invest.,* 74, 715, 1984.
3. **Le, P. H., Leiter, E. H., and Leyendecker, J. R.,** Genetic control of susceptibility to streptozotocin diabetes in inbred mice: effect of testosterone and H-2 haplotype, *Endocrinology,* 116, 2450, 1985.
4. **Weiner, H. L., Friedman, A., Neller, A., Khoury, S. J., al-Sabbagh, A., Santos, L., Savegh, M., Nussenblatt, R. B., Trentham, D. E., and Haffer, D. A.,** Oral tolerance: immunologic mechanisms and treatment of animal and human organ specific autoimmune diseases by oral administration of autoantigens, *Ann. Rev. Immunol.,* 12, 809, 1994.
5. **Epplen, J. T.,** On genetic components of autoimmunity: a critical review based on evolutionarily oriented rationality, *Hum. Genet.,* 90, 931, 1992.
6. **Deschampes, I. and Khalil, I.,** The role of the HLA system in the genetics of Type I Diabetes, *Diabetes Metab.,* 18, 253, 1992.
7. **Marchalonis, J. J., Schluter, S. F., Wang, E., Dehghanpeshek, K., Lake, D., Yocum, D. E., Edmundson, A. B., and Winfield, J. B.,** Synthetic autoantigens of immunoglobulins and T cell receptors: their recognition in aging, infection, and autoimmunity, *Proc. Soc. Exp. Biol. Med.,* 207, 129, 1994.
8. **Karounos, D. G. and Thomas, J. W.,** Recognition of common islet antigen by autoantibodies from NOD mice and humans with IDDM, *Diabetes,* 39, 1085, 1990.
9. **Berdanier, C. D.,** Diet, autoimmunity and insulin dependent diabetes mellitus: a controversy, *Proc. Soc. Exp. Biol. Med.,* 209, 223, 1995.
10. **Bellegrai, D., Naji, A., Silvers, W. K., Markmann, J. F., and Barker, C. F.,** Spontaneous diabetes in BB rats: evidence for a T cell dependent immune response defect, *Diabetologia,* 23, 359, 1982.
11. **Jackson, R., Rassi, N., Crump, T., Haynes, B., and Eisenbarth, G. S.,** The BB diabetic rat: profound T cell lymphocytopenia, *Diabetes,* 30, 887, 1981.
12. **MacKay, P., Jackson, J., and Rabinovitch, A.,** Spontaneous diabetes mellitus in the BB rat, *J. Clin. Invest.,* 77, 916, 1986.
13. **Like, A. A., Butler, L., Williams, R. M., Appel, M. C., Weringer, E. J., and Rossini, A.,** Spontaneous autoimune diabetes in the BB rat, *Diabetes,* 31, 7, 1982.
14. **Shimizu, J., Ikehara, S., Toki, J., Ohtsuki, H., Yasumizu, R., and Hamashima, Y.,** Immunohistochemical and ultrastructural studies on b cell destruction in NOD mouse, an animal model for type I diabetes mellitus, *Acta Histochem. Cytochem.,* 20, 9, 1987.
15. **Yale, J. F.,** Cyclosporin A for prevention and therapy of Type I diabetes in the BB rat, in *Frontiers in Diabetes Research. Lessons from Animal Diabetes II,* John Libby, London, 1988, 145.
16. **Laupacis, A., Steller, R. C., Gardell, C., Dupre, J., Keown, P., and Wallace, A. C.,** Cyclosporin prevents diabetes in BB Wistar rats, *Lancet,* 1, 10, 1983.
17. **Stiler, C. R., Dupre, J., Gent, M., Jenner, M. R., Keown, P. A., Laupacis, A., Martell, R., Rodger, N. W., Graffenreid, B. V., and Wolfe, B. M. J.,** Effects of cyclosporin imunosuppression in insulin dependent diabetes of recent onset, *Science,* 223, 1363, 1984.
18. **Baldari, C. T., Macchia, G., Heguy, A., Melli, M., and Telford, J. L.,** Cyclosporin A blocks calcium-dependent pathway of gene activation, *J. Biol. Chem.,* 266, 1910, 1991.
19. **Craignead, J. E.,** Workshop on viral infection and diabetes mellitus in man, *J. Infect. Dis.,* 215, 568, 1972.
20. **Craighead, J. E.,** Animal model of human disease, *Am. J. Path.,* 78, 537, 1975.
21. **Portman, N. D. and Taylor, K. W.,** Coxsackie B-4 virus-induced changes in mouse panreatic b cell mRNAs, *Biochem. Soc. Trans.,* 18, 1264, 1990.

22. **Leiter, E. H. and Wilson, G. L.,** Viral interactions in pancreatic b cells, in *Pathology of the Endocrine Pancreas,* Ripeleers, D. and Lefebore, P., Eds., Springer Verlag, Berlin, 1988, 8.
23. **Portwood, N. D. and Taylor, K.,** Coxsackie B4 virus-induced changes in mouse pancreatic b cell in mRNA's, *Biochem. Soc. Trans.,* 18, 1264, 1990.
24. **Notkins, A. L., Yoon, J. W., Onodera, T., Toniolo, A., and Jenson, A. B.,** Virus-induced diabetes mellitus, in *Perspectives in Virology XI,* Pollard, M., Ed., Alan R. Liss, New York, 1981, 141.
25. **Toniolo, A., Onodera, T., Jordan, B. G., Yoon, J-W., and Notkins, A. L.,** Virus induced diabetes mellitus. Glucose abnormalities produced in mice by the six members of the Coxsackie B virus group, *Diabetes,* 31, 496, 1982.
26. **Yoon, J-W., Austin, M., Onodera, T., and Notkins, A. L.,** Virus induced diabetes mellitus. Isolation of a virus from the pancreas of a child with diabetic ketoacidosis, *N. Eng. J. Med.,* 300, 1173, 1979.
27. **Hayashi, K., Boucher, D. W., and Notkins, A. L.,** Virus induced diabetes mellitus. II. Relationship between b cell damage and hyperglycemia in mice infected with encephalomyocarditis virus, *Am. J. Path.,* 75, 91, 1982.
28. **Jensen, B. A., Rosenberg, H. S., and Notkins, A. L.,** Postmortem tissue changes in children with fatal viral infections, *Lancet,* 2, 354, 1980.
29. **Babu, P. G., Huber, S. A., and Craighead, J. E.,** Immunology of viral diabetes, *Surv. Synth. Path. Res.,* 4, 1, 1985.
30. **Loria, R. M., Montgomery, L. B., Tuttle-Fuller, N., Gregg, H. M., and Chinchilli, V. M.,** Genetic predisposition to diabetes mellitus is assoiated with impaired humoral immunity to Coxsackie virus B4, *Diabetes Res. Clin. Prac.,* 2, 91, 1986.
31. **Bae, Y-S., Eun, H-M., and Yoon, J-W.,** Molecular identification of diabetic viral gene, *Diabetes,* 38, 316, 1989.
32. **Szopa, T. M., Ward, T., Dronfield, D. M., Portwood, N. D., and Taylor, K. W.,** Coxsackie B4 viruses with the potential to damage b cells of the islets are present in clinical isolates, *Diabetologia,* 33, 325, 1990.
33. **Shafrir, E. and Renold, A. E., Eds.,** *Lessons in Animal Diabtes I–III,* Smith Gordon, London, 1992.
34. **Melloul, D. and Cerasi, E.,** Transcription of the insulin gene: towards defining the glucose sensitive cis-element and trans-acting factors, *Diabetologia,* 37, 53, 1994.
35. **Steiner, D. F., Tager, H. S., Chan, S. J., Nanjo, K., Sanke, T., and Rubenstein, A. H.,** Lessons learned from molecular biology of insulin gene mutations, *Diabetes Care,* 13, 600, 1990.
36. **Olansky, L., Janssen, R., Welling, C., and Permutt, M. A.,** Variability of the insulin gene in American Blacks with NIDDM, *Diabetes,* 41, 742, 1992.
37. **Naggert, J. K., Fricker, L. D., Varlamov, O., Nishima, P. M., Rouille, Y., Steiner, D. F., Carroll, R. J., Paigen, B. J., and Leiter, E. H.,** Hyperproinsulinemia in obese fat/fat mice associated with a carboxypeptidase E mutation which reduces enzyme activity, *Nat. Genet.,* 10, 135, 1995.
38. **Stoffel, M., Froguel, P. H., Takeda, J., Zouali, H., Vionnet, N., Nishi, S., Weber, I. T., Harrison, R. W., Pilkis, S. J., Lesage, S., Vaxillaire, M., Velho, G., Sun, F., Iris, F., Passa, P. H., Cohen, D., and Bell, G. I.,** Human glucokinase gene: isolation, characterization, and identificaton of two missense mutations linked to early onset-non insulin-dependent (type 2) diabetes mellitus, *Proc. Natl. Acad. Sci. U.S.A.,* 89, 7689, 1992.
39. **Rowe, R. E., Wapelhorst, B., Bell, G. I., Risch, N., Spielman, R. S., and Concannon, P.,** Linkage and association between insulin-dependent diabetes mellitus susceptibility and markers near glucokinase gene on chromosome 7, *Nat. Genet.,* 10, 240, 1995.
40. **Zouali, H., Vaxillaire, M., Lesage, S., Sun, F., Velho, G., Vionnet, N., Chiu, K., Passa, P., Permutt, A., Demenais, F., Cohen, D., Beckmann, J. S., and Froguel, P.,** Linkage analysis and molecular scanning of glucokinase gene in NIDDM families, *Diabetes,* 42, 1238, 1993.

41. **Caro, J. F., Triester, S., Patel, V. K., Tapscott, E. B., Frazier, N. L., and Dohm, G. L.,** Liver glucokinase decreased activity in patients with Type II diabetes, *Horm. Metab. Res.,* 27, 19, 1995.
42. **Magnuson, M. A. and Jetton, T. L.,** Tissue specific regulation of glucokinase, in *Nutrition and Gene Expression,* Berdanier, C. D. and Hargrove, J. L., Eds., CRC Press, Boca Raton, FL, 1993, 143.
43. **Erecinska, M., Bryla, J., Michalik, M., Meglason, M. D., and Nelson, D.,** Energy metabolism in islets of Langerhans, *Biochem. Biophys. Acta,* 1101, 273, 1992.
44. **Robertson, R. P.,** Defective insulin secretion in NIDDM: integral part of a multiplier hypothesis, *J. Cell. Biochem.,* 48, 227, 1992.
45. **Tsaur, M-L., Menzel, S., Lai, F-P., Espinosa, R., Concannon, P., Spielman, R. S., Harris, C. L., Cox, N. J., LeBeau, M. M., German, M. S., Jan, L. Y., Bell, G. I., and Stoffel, M.,** Isolation of a cDNA clone encoding a K_{ATP} channel-like protein expressed in insulin secreting cells, localization of the human gene to chromosome band 21q22.1 and linkage studies with NIDDM, *Diabetes,* 44, 592, 1995.
46. **Bjorntorp, P. and Brodoff, B. N., Eds.,** *Obesity,* J. B. Lippincott Co., Philadelphia, 1962.
47. **Makino, H., Taire, M., Shimada, F., Hasimoto, N., Suzuki, Y., Nozaki, O., Hatanaka, Y., and Yoshida, S.,** Insulin receptor gene mutation: a molecular and functional analysis, *Cellular Signaling,* 4, 351, 1992.
48. **Ebina, Y., Ellis, L., Jarnagin, K., Edery, M., Graf, L., Clauser, E., Ou, J-H., Masiarz, F., Kan, Y. W., Goldfine, I. D., Roth, R. A., and Rutter, W. J.,** The insulin receptor cDNA: the structural basis for hormone-activated transmembrane signalling, *Cell,* 40, 747, 1985.
49. **Ullrich, A., Bell, J. R., Chen, E. Y., Herrera, Y., Petruzzelli, L. M., Dull, T. J., Gray, A., Cousins, L., Liao, Y-C., Tsubokawa, M., Mason, H., Seeburg, P. H., Grunfeld, C., Rosen, O. M., and Ramachandran, J.,** Human insulin receptor and its relationship to the tyrosine kinase, *Nature,* 313, 756, 1985.
50. **Levy, J. R. and Hug, V.,** Regulation of insulin gene expression. Cell cycle mediated effects on insulin receptor mRNA stability, *J. Biol. Chem.,* 267, 25289, 1992.
51. **Baily, D. L. and Horuk, R.,** The biology and biochemistry of the glucose transporter, *Biochem. Biophys. Acta,* 947, 571, 1988.
52. **Gould, G. W. and Bell, G. I.,** Facilitative glucose transporters: an expanding family, *TIBS,* 15, 18, 1990.
53. **Kahn, B. B. and Pederson, O.,** Tissue specific regulation of glucose transporters in different forms of obesity, *Proc. Soc. Exp. Biol. Med.,* 200, 214, 1992.
54. **Seino, Y., Yuamamoto, T., and Koh, G.,** Insulin and glucose transporter gene expression in obesity and diabetes, *Proc. Soc. Exp. Biol. Med.,* 200, 210, 1992.
55. **Joost, H. G. and Weber, T. M.,** Regulation of glucose transport in insulin sensitive cells, *Diabetologia,* 32, 831, 1989.
56. **Orci, L., Ravazzola, M., Baetens, D., Inman, L., Amherdt, M., Peterson, R. G., Newgard, C. B., Johnson, J. H., and Unger, R. H.,** Evidence that down regulation of b cell glucose transporters in NIDDM may be the cause of diabetic hyperglycemia, *Proc. Natl. Acad. Sci. U.S.A.,* 87, 9953, 1990.
57. **Sleiker, L. J., Sundell, K. L., Heath, W. F., Osborne, H. E., Bue, J., Manetta, J., and Sportsman, J. R.,** Glucose transporter levels in tissues of spontaneously diabetic Zucker fa/fa rat (ZDF/drt) and viable yellow mouse (A^{vy}/a), *Diabetes,* 41, 187, 1992.
58. **Pederson, O., Kahn, R., Flier, J. S., and Kahn, B. B.,** High fat feeding causes insulin resistance and a marked decrease in the expression of GLUT 4 transporters in fat cells of rats, *Endocrinology,* 129, 771, 1991.
59. **Vranic, M.,** Glucose turnover. A key to understanding the pathogenesis of diabetes. (Indirect effects of Insulin), *Diabetes,* 41, 1188, 1992.
60. **Barnard, R. J. and Youngren, J. F.,** Regulation of glucose transport in skeletal muscle, *FASEB J.,* 6, 3238, 1992.
61. **Mitchell, P.,** Coupling of phosphorylation to electron and hydrogen transfer by a chemi-osmotic type of mechanism, *Nature,* 191, 144, 1961.

62. **Schwerzmann, K. and Pedersen, P. L.,** Regulation of the mitochondrial ATP synthase/ATPase complex, *Arch. Biochem. Biophys.,* 250, 1, 1986.
63. **Nicholls, D. G. and Locke, R. M.,** Thermogenic mechanisms in brown fat, *Physiol. Rev.,* 64, 1, 1984.
64. **Sundin, U. and Cannon, B.,** GDP-binding to the brown fat mitochondria of developing and cold-adapted rats, *Comp. Biochem.,* 65B, 460, 1980.
65. **Kopecky, J., Juerri, F., Jezek, P., Drahota, Z., and Houstek, J.,** Molecular mechanism of uncoupling in brown adipose tissue mitochondria. The non-identity of proton and chloride conducting pathways, *FEBS Lett.,* 170, 186, 1984.
66. **Kozak, L. P., Britton, J. H., Kozak, U. C., and Wells, J. M.,** The mitochondrial uncoupling protein gene. Correlation of exon structure to transmembrane domains, *J. Biol. Chem.,* 263, 12274, 1988.
67. **Himms-Hagen, J.,** Brown adipose tissue metabolism and thermogenesis, *Ann. Rev. Nutr.,* 5, 69, 1985.
68. **Trayhurn, P.,** Brown fat tissue and the regulation of energy balance, in *Brown Adipose Tissue,* Trayhurn, P. and Nicholls, D. G., Eds., Edward Arnold, London, 1986, 299.
69. **Otto, D. A. and Ontko, J. A.,** Structure-function relations between fatty acid oxidation and the mitochondrial inner-membrane matrix region, *Eur. J. Biochem.,* 129, 479, 1982.
70. **Williams, M. A., Stancliff, R. C., Packer, L., and Keith, A. D.,** Relation of unsaturated fatty acid composition of rat liver mitochondria to oscillation period, spin label motion, permeability and oxidative phosphorylation, *Biochim. Biophys. Acta,* 267, 444, 1972.
71. **Smith, J. A. and DeLuca, H.,** Effect of dietary fats on oxidative phosphorylation and fatty acid profile of rat liver mitochondria, *J. Cell Biol.,* 21, 15, 1964.
72. **Divakaran, P. and Venkataraman, A.,** Effect of dietary fats on oxidative phosphorylation and fatty acid profile in rat liver mitochondria, *J. Nutr.,* 107, 1621, 1977.
73. **Robblee, N. M. and Clandinin, M. T.,** Effect of dietary fat level and polyunsaturated fatty acid content on the phospholipid composition of rat cardiac mitochondrial membranes and mitochondrial ATPase activity, *J. Nutr.,* 114, 263, 1984.
74. **Abuirmeileh, N. M. and Elson, C. E.,** The influence of linoleic acid intake on membrane bound respiratory activities, *Lipids,* 15, 918, 1980.
75. **Rafael, J., Patzelt, J., Schafer, H., and Elmadfa, I.,** The effect of essential fatty acid deficiency on basal respiration and function of liver mitochondria in rats, *J. Nutr.,* 114, 255, 1984.
76. **Deaver, O. E., Jr., Wander, R. C., McCusker, R. H., and Berdanier, C. D.,** Diet effects on membrane phospholipid fatty acids and mitochondrial function in BHE rats, *J. Nutr.,* 116, 1148, 1986.
77. **Chance, B. and Williams, J.,** A simple and rapid assay of oxidative phosphorylation, *Nature,* 175, 1120, 1955.
78. **Reardon, W., Roso, R. J. M., Sweeney, M. G., Luxon, L. M., Pembrey, M. E., Harding, A. E., and Trembath, R. C.,** Diabetes mellitus associated with a pathogenic point mutation in mtDNA, *Lancet,* 340, 1376, 1992.
79. **Gerbitz, K. D., Paprotta, A., Jaksch, M., Zienz, S., and Dreschel, J.,** Diabetes mellitus is one of the heterogeneous phenotypic features of a mtDNA point mutation within the $tRNA^{Leu(UUR)}$ gene, *FEBS Lett.,* 321, 194, 1993.
80. **Van den Ouweland, J. M. W., Lemkes, H. H. P. J., Trembath, R. C., Ross, R., Velho, G., Cohen, D., Froguel, P., and Maassen, J. A.,** Maternally inherited diabetes and deafness is a distinct subtype of diabetes and associates with a single point mutation in the mitochondrial $tRNA^{Leu(UUR)}$ gene, *Diabetes,* 43, 746, 1994.
81. **Kishimoto, M., Hashiramoto, M., Araki, S., Ishida, Y., Kazumi, T., Kanda, F., and Kasuga, M.,** Diabetes mellitus carrying a mutation in the mitochondrial $tRNA^{Leu(UUR)}$ gene, *Diabetologia,* 38, 193, 1995.
82. **Moraes, C. T., Ricci, E., Bonilla, E., DiMauro, S., and Schon, E. A.,** The mitochondrial $tRNA^{Leu(UUR)}$ mutation in mitochondrial encephalomyopathy, lactic acidosis and stroke like episodes (MELAS): genetic, biochemical and morphological correlations in skeletal muscle, *Am. J. Hum. Genet.,* 50, 934, 1992.

83. **Nakagawa, Y., Ikegami, H., Yamato, E., Takekawa, K., Fujisawa, T., Hamada, Y., Ueda, H., Uchigata, Y., Mike, T., Kumahara, Y., and Ogihara, T.,** A new mtDNA mutation associated with non insulin dependent diabetes mellitus, *Biochem. Biophys. Res. Commun.,* 209, 664, 1995.
84. **Ballinger, S. W., Shoffner, J. M., Hedaya, E. V., Trounce, I., Polak, M. A., Koontz, D. A., and Wallace, D. C.,** Maternally transmitted diabetes and deafness associated with a 10.4 kb mtDNA deletion, *Nat. Genet.,* 1, 11, 1992.
85. **Rotig, A., Bessis, J-L., Romero, N., Cormier, V., Saudubray, J-M., Narcy, P., Lenoir, G., Rustin, P., and Munnich, A.,** Maternally inherited duplication of the mitochondrial genome in a syndrome of proximal tubulopathy, diabetes mellitus, and cerebellar ataxia, *Am. J. Hum. Genet.,* 50, 364, 1992.
86. **Berdanier, C. D.,** NIDDM in the nonobese BHE/cdb rat, in *Recent Advances in Animal Diabetes,* Shafrir, E., Ed., Smith Gordon Publishers Ltd., London, 1994, 231.
87. **Lakshmanan, M. R., Berdanier, C. D., and Veech, R. L.,** Comparative studies on lipogenesis and cholesterogenesis in lipemic BHE rats and normal Wistar rats, *Arch. Biochem. Biophys.,* 183, 355, 1977.
88. **Berdanier, C. D.,** Rat strain differences in gluconeogenesis in isolated hepatocytes, *Proc. Soc. Exp. Biol. Med.,* 169, 74, 1982.
89. **Liang, Y., Bonner-Weir, S., Wu, Y-J., Berdanier, C. D., Berner, D. K., Efrat, S., and Matschinsky, F.,** In situ glucose uptake and glucokinase activity of pancreatic islets in diabetic and obese rodents, *J. Clin. Invest.,* 93, 2473, 1994.
90. **Berdanier, C. D., Johnson, B., Hartle, D. K., and Crowell, W.,** Lifespan is shortened in BHE/cdb rats fed a diet containing 9% menhaden oil and 1% corn oil, *J. Nutr.,* 122, 1309, 1992.
91. **Fowler, K. A., Crowell, W. A., and Berdanier, C. D.,** Early renal disease in BHE/cdb rats is less in rats fed beef tallow than in rats fed menhaden oil, *Proc. Soc. Exp. Biol. Med.,* 203, 163, 1993.
92. **Bouillon, D. J. and Berdanier, C. D.,** Effects of maternal carbohydrate on mitochondrial activity and on lipogenesis by the young and mature progeny, *J. Nutr.,* 113, 2205, 1983.
93. **McCusker, R. H., Deaver, O. E., and Berdanier, C. D.,** Effect of sucrose or starch feeding on the hepatic mitochondrial activity of BHE and Wistar rats, *J. Nutr.,* 113, 1327, 1983.
94. **Berdanier, C. D., Tobin, R. B., and DeVore, V.,** Studies on the control of lipogenesis: strain differences in hepatic mitochondria, *J. Nutr.,* 109, 247, 1979.
95. **Berdanier, C. D. and MacNamara, S.,** Aging and mitochondrial activity in BHE and Wistar rats, *Exp. Gerontol.,* 15, 519, 1980.
96. **Berdanier, C. D. and Thomson, A.,** Comparative studies on mitochondrial respiration in four strains of rats, *Comp. Biochem. Physiol.,* 85B, 531, 1986.
97. **Berdanier, C. D. and Kim, M. C.,** Hyperthyroidism does not induce an increase in mitochondrial respiration, *J. Nutr. Biochem.,* 4, 10, 1993.
98. **Mathews, C. E., McGraw, R. A., and Berdanier, C. D.,** A point mutation in the mitochondrial DNA of diabetes-prone BHE/cdb rats, *FASEB J.,* 9, 1638, 1995.
99. **Grosskopf, R. and Feldmann, H.,** Analysis of a DNA segment from rat liver mitochondria containing the genes for the cytochrome oxidase subunits I, II, III and ATPase subunit 6 and several tRNA genes, *Curr. Genet.,* 4, 151, 1981.
100. **Lang, P. L.,** Crl:CD®BR Rat; A collection of Charles River Lab Publications, Charles River WIGA, Deutschland, Sulzfeld.
101. **Gadalata, G., Pepe, G., DeCandia, G., Quagliariello, C., Sbissa, E., and Sacone, C.,** The complete nucleotide sequence of *Rattus norvigicus* mitochondrial genome: cryptic signals revealed by comparative analysis between vertebrates, *J. Mol. Evol.,* 28, 497, 1989.
102. **Brown, G. C.,** The leaks and slips of bioenergetic membranes, *FASEB J.,* 6, 2961, 1992.
103. **Brown, G. C., Lakin-Thomas, P. L., and Brand, M. D.,** Control of respiration and oxidative phosphorylation in isolated rat liver cells, *Eur. J. Biochem.,* 192, 355, 1990.
104. **Hafner, R. P., Brown, G. C., and Brand, M. D.,** Analysis of the control of respiration rate, phosphorylation rate, proton leak rate and protomotive force in isolated mitochondria using the "top down" approach to metabolic control theory, *Eur. J. Biochem.,* 188, 313, 1990.

105. **Brown, G. C.,** Control of respiration and ATP synthesis in mammalian mitochondria and cells, *Biochem. J.*, 284, 1, 1992.
106. **Boyer, P. D.,** A perspective of the binding charge mechanism for ATP synthesis, *FASEB J.*, 3, 2164, 1989.

Chapter 9

MITOCHONDRIAL DNA MUTATIONS AND LIPID PEROXIDATION IN HUMAN AGING

Yau-Huei Wei and Shu-Huei Kao

CONTENTS

0-8493-9485-6/96/$0.00+$.50

INTRODUCTION

Aging has been recognized as a major risk factor of human diseases and death.[1] It is accompanied by a progressive decline in the biochemical and physiological functions of various organs in an individual. Most theories of aging have centered around the notion that age-related deterioration of physiological functions is primarily due to functional decline caused by qualitative and quantitative modifications of cellular constituents. Recently, various lines of evidence have been gathered to support the free-radical theory of aging as proposed more than four decades ago.[2] Although cells have developed various enzymatic and nonenzymatic systems to protect them from damage by reactive oxygen species (ROS),[3] a fraction of ROS may escape the cellular defense mechanims and cause permanent or transient damage to proteins, lipids, and nucleic acids.[4-6] The ROS are also thought to be involved in the pathogenesis of many common disorders during human aging including atherosclerosis, cancer, ischemia, chronic inflammation, arthritis, and the autoimmune diseases.[4-9] The free-radical attack of mitochondrial DNA (mtDNA) could be involved in these pathological conditions because of the fact that mitochondria are the major source of ROS and free radicals in cells. There are several reasons why this occurs: the mtDNA is naked (has no protective histone coat), compact, and is preferentially attached to the mitochondrial inner membrane, which is the major site of lipid peroxidation in the mitochondria.

With the advances of analytical techniques in molecular biology, more than two dozen large-scale deletions have been recently reported to occur in the mtDNA of various tissues of old humans (Figure 1).[10-15] In addition, two disease-associated point mutations in $tRNA^{Lys}$ and $tRNA^{Leu(UUR)}$ genes, respectively, in mtDNA were also detected in the muscle of aged individuals.[16,17] In the past few years, we have demonstrated that liver mitochondrial respiratory function declines with age[18] and that there are at least three types of aging-associated mtDNA deletions (5 kb, 6 kb, and 7.4 kb) in the livers of the elderly subjects.[10,13,14,19] In a further study, we showed that lipid peroxides are concomitantly accumulated in human liver mitochondria during aging.[20] In addition, we have employed a semi-quantitative polymerase chain reaction (PCR) method[19] to determine the amount of mutant mtDNAs, including 4,977-bp deletion, 7,436-bp deletion, and D-loop tandem duplications, and compared the proportions of the mutant mtDNAs in the muscle, liver, and testis from human subjects of different ages. On the other hand, we used the high performance liquid chromatographic (HPLC) techniques to determine the amount of lipid peroxides in the mitochondria of various human tissues. These studies have enabled us to examine whether aging is associated with a simultaneous increase in mitochondrial lipid peroxidation and mtDNA mutations, and whether there is a correlation between the two events in various tissues of the human in the aging process. The association of the 4,977-bp and 7,436-bp deletions and D-loop tandem duplications of mtDNA and lipid peroxidations with human aging is reviewed and some pertinent issues are discussed.

THE GENETICS OF MITOCHONDRIAL DNA

Mitochondria are the intracellular organelles that are responsible for ATP synthesis through the coupling of oxidative phosphorylation (OXPHOS) to mitochondrial respiration. The capacity of energy metabolism of mitochondria has been demonstrated to decline with age in various tissues.[18,21-23] Mitochondria contain their own genetic materials, which are the only extrachromosomal DNA in human cells.[24,25] Human mtDNA is a maternally inherited, 16,569-bp circular double-stranded DNA, which encodes 13 polypeptides constituting the respiratory enzyme complexes, 2 rRNAs, and 22 tRNAs.[24] Recently, Linnane et al.[26] hypothesized that accumulation of somatic mutations in the mtDNA is a major contributor to human aging and degenerative diseases. MtDNA is especially vulnerable to oxidative damage because it lacks histones and other DNA-binding proteins and replicates rapidly without proofreading and efficient DNA repair mechanisms.[27,28] These characteristics of mtDNA have rendered it vulnerable to mutation at a rate 17 times higher than the nuclear DNA.[29] Although many of the mutations are silent or associated with genetic polymorphism, there are some missense or nonsense mutations in the structural genes or tRNA genes in mtDNA that may have pathological consequences.[15,26] Over the past five years, more than two dozen large-scale deletions in the mtDNA of various tissues of old humans (Figure 1) have been reported by several investigators. Some point mutations in the tRNA genes[16,17] and tandem duplications in the D-loop region of mtDNA[30] have also been demonstrated to be associated with human aging. We believe that mtDNA mutations are early molecular events associated with human aging and are responsible for the age-dependent decline of mitochondrial respiratory functions in various human tissues.[10,21]

MITOCHONDRIAL DNA MUTATIONS IN HUMAN AGING

Significant progress has been made in the study of the age-related somatic mtDNA mutations in human tissues. MtDNA has a very high mutation rate, involving both nucleotide substitutions and insertion-deletion mutations. Mutations that occur in the germ line cells result in familial diseases, while those that occur during development or in somatic tissues may cause sporadic diseases and age-related decline in the function of the OXPHOS system.[9,15,31] Pathological mutations of mtDNA fall into two major classes: deletion-insertion and point mutation. Deletions in mtDNA have been reported that either spare both origins of replication, or remove the origin of replication of the light strand (OL).[32,34] Insertion can cause duplication of gene sequences in various mtDNA regions. The base substitution mutations that have been identified include both missense (amino acid substitution in structural genes) and protein synthesis (tRNA and rRNA) gene mutations (Table 1). Thus, the accumulation of somatic mtDNA mutations could contribute to the progression of mitochondrial

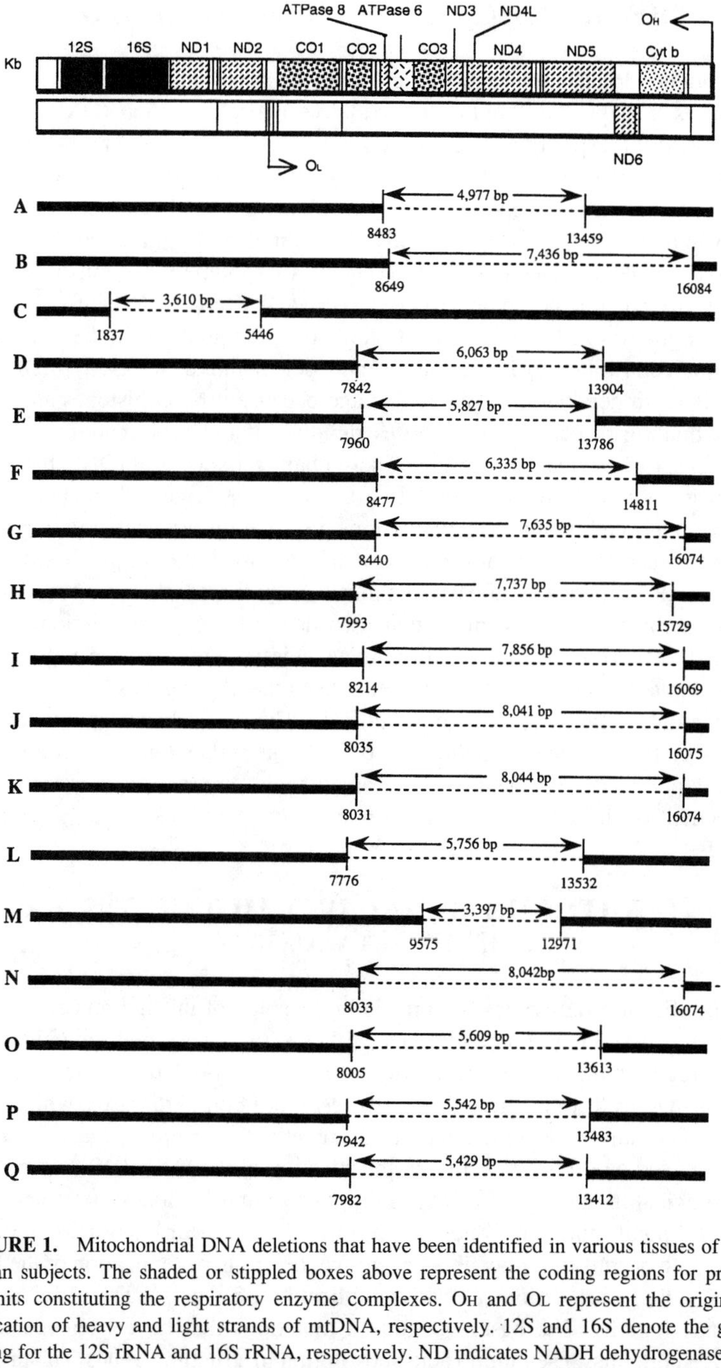

FIGURE 1. Mitochondrial DNA deletions that have been identified in various tissues of aged human subjects. The shaded or stippled boxes above represent the coding regions for protein subunits constituting the respiratory enzyme complexes. OH and OL represent the origins of replication of heavy and light strands of mtDNA, respectively. 12S and 16S denote the genes coding for the 12S rRNA and 16S rRNA, respectively. ND indicates NADH dehydrogenase, CO indicates cytochrome oxidase, Cyt *b* indicates cytochrome *b,* and ATPase represents Fo,F1-ATPase. From this figure one can easily identify the genes that are truncated or removed by each type of deletion.

TABLE 1
mtDNA Mutations Associated with Mitochondrial Diseases Other than LHON

Type of Mutation	Mutated Genes	Clinical Diagnosis	Nucleotide Position (Substitution)	Amino Acid or RNA	Homo-plasmy	Hetero-plasmy
Missense	ATPase 6	NARP	8993 (T to G)	L to R	—	+
	ATPase 6	NARP	8993 (T to C)	L to P	—	+
	ND 4	MELAS	11084 (A to G)	R to H	+	+
tRNA	tRNALeu	MELAS	3243 (A to G)	tRNA$^{Leu(UUR)}$	—	+
	tRNA^{Leu}	DMDF	3243 (A to G)	tRNA$^{Leu(UUR)}$	—	+
	tRNA^{Leu}	MM	3250 (T to C)	tRNA$^{Leu(UUR)}$	—	+
	tRNA^{Leu}	MMC	3260 (A to G)	tRNA$^{Leu(UUR)}$	—	+
	tRNA^{Leu}	MELAS	3271 (T to C)	tRNA$^{Leu(UUR)}$	—	+
	tRNA^{Leu}	MM	3302 (A to G)	tRNA$^{Leu(UUR)}$	—	+
	tRNA^{Leu}	MMC	3303 (C to T)	tRNA$^{Leu(UUR)}$	—	+
	tRNA^{Gln}	ADPD	4336 (A to G)	tRNA^{Gln}	+	—
	tRNA^{Lys}	MERRF	8344 (A to G)	tRNA^{Lys}	—	+
	tRNA^{Lys}	MERRF	8356 (T to C)	tRNA^{Lys}	—	+
	tRNA^{Lys}	MERRF/MELAS	8356 (T to C)	tRNA^{Lys}	—	+
	tRNA^{Gly}	MHCM	9997 (T to C)	tRNA^{Gly}	?	+
	tRNA^{Gly}	CIPO	10006 (A to G)	tRNA^{Gly}	?	?
	tRNA^{Ser}	CIPO	12246 (C to G)	tRNA^{Ser}	?	?
	tRNA^{Leu}	CIPO	12308 (A to G)	tRNA^{Leu}	?	?
	tRNA^{Leu}	CPEO	12308 (A to G)	tRNA^{Leu}	?	?
	tRNA^{Thr}	LIMM	15923 (A to G)	tRNA^{Thr}	+	—
	tRNA^{Pro}	MM	15990 (G to A)	tRNA^{Pro}	—	+
rRNA	12S rRNA	DEAF	1555 (A to G)	12S rRNA	+	—
	16S rRNA	ADPD	3196 (G to A)	16S rRNA	+	+

Abbreviations: ADPD, Alzheimer's disease and Parkinson's disease; CIPO, chronic intestinal pseudoobstruction with myopathy and opthalmoplegia; CPEO, chronic progressive external ophthalmoplegia; DEAF, maternally inherited deafness or aminoglycoside-induced deafness; DMDF, diabetes mellitus and deafness; LIMM, lethal infantile mitochondrial myopathy; MHCM, maternally inherited hypertrophic cardiomyopathy; NARP, neurogenic muscle weakness, ataxia, and retinitis pigmentosa; MM, mitochondrial myopathy; MMC, maternally inherited mitochondrial myopathy and cardiomyopathy.

diseases, the occurrence of various types of degenerative diseases, and the age-related decline in the biochemical function of the OXPHOS system.[10,15]

DELETION-INSERTION MUTATIONS

Deletions

Deletions of mtDNA were first identified in 1988 in patients with chronic progressive external ophthalmoplegia (CPEO),[32] with or without myopathy, and in patients with Kearns-Sayre syndrome (KSS).[33] Patients with mtDNA deletions usually appear as sporadic cases without a family history of mitochondrial diseases. The proportion of the deleted mtDNAs in the target tissue may increase with age.[33] Analysis of the respiratory chain function in the skeletal muscle from patients with mtDNA deletions often reveals multiple deficiencies involving Complex I, III, and IV.[34] MtDNA deletions are heteroplasmic and are most commonly found in clinically affected tissues of patients with CPEO or KSS.[32-34] MtDNA deletions vary in size, although they are often constant within the tissues of an individual. A 4,977-bp deletion, termed the "5-kb common deletion," is most frequently seen in CPEO or KSS patients. Multiple mtDNA deletions have been found in autosomal dominant mitochondrial myopathies and inherited recurrent myoglobinuria.[35] The mechanisms by which mtDNA deletions cause diseases are still important issues of current research in the field. This kind of large-scale deletion often causes the removal or truncation of multiple structural genes and tRNA genes and so would be expected to cause multiple respiratory chain deficiencies that could be revealed by biochemical assays of respiratory enzyme functions.[34] The effect of the deletion of multiple tRNAs may be attenuated if complementation occurs between normal and deleted genomes within the mitochondria. Analysis of the breakpoints flanking the 4,977-bp deletion revealed that the deletion site is flanked by a 13-bp direct repeat (5′-ACCTCCCTCACCA-3′) at nucleotide positions (np) 8470/8482 and np 13447/13459, respectively. Several mechanisms have been proposed to explain the generation of the deletions including homologous recombination,[36] topoisomerase cleavage,[37] and slip-replication,[36] respectively. The other type of common deletion is "7,436-bp deletion," whose breakpoints are flanked by 12-bp direct repeats (5′-CATCAACAACCG-3′) at np 8637/8648 and np 16073/16084 of mtDNA, respectively. Deletions of 4,977 bp, 7,436 bp and several other large-scale deletions in the mtDNA of various tissues of old humans (Figure 1)[11-13,15] have been independently demonstrated by several laboratories.[11]

Insertions and Duplications of mtDNA

Duplications of mtDNA have been observed less frequently than the large-scale deletions. Three patients, one with KSS[38] and one with Pearson's syndrome,[39] and one with an inherited proximal tubulopathy, diabetes mellitus, and cerebral ataxia[40] have been reported. All three had duplications which were approximately 8 kb in length. A third family had a duplication of mtDNA sequence of about 10 kb in length,[41] detected by southern hybridization, in the muscle and blood cells of two siblings, and by PCR of the muscle mtDNA of the mother.[40] The clinical manifestations of patients in these three families were similar to those found in patients with mtDNA deletions. However, the pathophysiology of these duplications of mtDNA is not clear. The gene products made from the duplicated mtDNA sequences could be harmful, the abnormal polypeptides or transcripts might impair mitochondrial function by competing with normal gene products, and thereby cause the mitochondrial diseases. Recently, Brokington et al.[42] reported a heteroplasmic tandem duplication with an approximate size of 260 bp in the D-loop of human mtDNA and proposed that it is associated with large-scale deletions in mtDNAs of patients with mitochondrial myopathies. They further suggested that the duplications of mtDNA might be responsible for mild mitochondrial dysfunction in muscle of the mothers of some patients, and is a predisposition to mtDNA deletions.[42,43] Recently, we have investigated the question of whether tandemly duplicated mtDNA exists in the tissues of aged individuals.[30] Interestingly, we identified two types of aging-associated tandemly duplicated mtDNAs.[30,44] Among them, type I was identical to the disease-associated 260-bp tandem duplication, while type II was a 200-bp partially duplicated mtDNA.[30] Only some of them were found to be associated with human aging.

POINT MUTATIONS OF MITOCHONDRIAL DNA

Specific missense point mutations have been demonstrated to be associated with several ophthalomological and neurological diseases.[15,45] The most notable ones are Leber's hereditary optic neuropathy (LHON), two mitochondrial encephalomyopathies mercifully abbreviated as "MELAS" (mitochondrial encephalopathy, lactic acidosis and stroke-like episodes) and "MERRF" (myoclonic epilepsy and ragged-red fibers) syndromes, and neurogenic muscle weakness, ataxia, and retinitis pigmentosa (NARP) (see Table 1). Numerous mtDNA mutations have been identified to be associated with LHON.[9,45] These mutations are all located in structural genes encoding polypeptides essential for mitochondrial OXPHOS function. The most common cause of LHON is a mutation at np 11778 of mtDNA. This G to A transition changes a highly conserved arginine to a histidine at the 340th amino acid of the ND4 subunit, which is a component of the large hydrophobic protein domain of Complex I. The second common cause of LHON is a G to A transition in the ND1 gene at np 3460 of mtDNA. These mutations may be homoplasmic or heteroplasmic.[15,45] A hetroplasmic A to G transition at np 3243 or a T to C transition at np 3271 in the $tRNA^{Leu(UUR)}$ gene has been found in 80 to 85% of the MELAS patients.[15] The mutation lies within the binding site for the so-called termination associated sequence (TAS), which is involved in controlling the relative abundance of two rRNAs and mRNAs.[24,46] *In vitro,* the point mutation reduces binding of the termination associated protein (factor) and results in a lower level of the shorter of the alternative transcripts (two rRNAs).[46] An A to G transition at np 8344 in the TyC loop of $tRNA^{Lys}$ gene is present in 85 to 90% of the patients with MERRF syndrome.[47] This mutation is expected to "bend" the $tRNA^{Lys}$ structure and should result in faulty lysine incorporation during protein synthesis. A T to G heteroplasmic mutation in the ATPase 6 gene at np 8993 of mtDNA has been demonstrated to be associated with NARP.[48] This mutation has also been identified in some of the patients with Leigh's disease.[49] This mutation causes the substitution of arginine for leucine and is believed to interfere with the charge distribution, proton entry, and ATP synthesis of F_0,F_1-ATPase of the mitochondria. Among the mutations that have been proved to be associated with specific human diseases, the A to G transitions at np 3243 and np 8344 of mtDNA have been demonstrated to occur also in aging human tissues.[16,17] However, some other point mutations in mtDNA have recently been found to be frequently associated with several age-related degenerative diseases such as Alzheimer's disease and Parkinson's disease.[15,50]

LENGTH MUTATIONS AND TANDEM DUPLICATIONS OF MTDNA ARE ACCUMULATED DURING AGING

AGE-DEPENDENT INCREASE OF THE DELETED mtDNAS IN HUMAN TISSUES

Over the past five years, we have employed PCR techniques to identify the large-scale deletions by using different primer pairs (Table 2). The sequences of the oligonucleotide primers and the sizes of PCR products amplified from the 4,977-bp and 7,436-bp deleted mtDNAs, respectively, using each of the primer pairs are shown in Table 2. The incidence of the 4,977-bp and 7,436-bp deleted mtDNAs in various human tissues from older subjects was higher than those from younger subjects (Table 3). However, the 4,977-bp deleted mtDNA was undetectable in tissues from the young subjects and in all of the blood samples from healthy subjects examined. In order to ascertain that amplified DNA fragment was not due to misannealing of primers, we identified the deletions of mtDNA by a primer-shift PCR method.[19] By this method, we have been able to conclude the presence of an authentic 4,977-bp deletion in mtDNA of aging human tissues or cells with defective bioenergetic functions (Figure 2A). By the same experimental design, we were able to demonstrate the presence of a 7,436-bp deletion in the mtDNA of old human tissues (Figure 2B).

ACCUMULATION OF THE DELETED mtDNAS IN OLD HUMAN TISSUES

In order to investigate the proportion of mtDNA molecules with a specific deletion in human individuals, we developed a semi-quantitative PCR method[19] to determine the relative amount of the mutant mtDNA in human tissues from subjects of different ages. We first determined the proportion of the 4,977-bp deleted mtDNA from the human specimens that had been confirmed to harbor the deletion. We found that the proportions of the deleted mtDNA in the muscle, liver, and testis samples were all increased with age (Figure 3). In addition, we investigated whether the 7,436-bp deleted mtDNA is also accumulated with age in human tissues. The results demonstrated that the proportion of the 7,436-bp deleted mtDNA was indeed increased with age (Figure 4). These finding are consistent with those obtained in the previous studies of the 4,977-bp deletion in various tissues of aged individuals.[11-14,19] By use of the semi-quantitative PCR method, we found that the proportions of the 4,977-bp and 7,436-bp deleted mtDNAs in normal human tissues including the muscle, liver, testis, and brain were in good agreement with the results obtained by

TABLE 2
Oligonucleotide Primers Used for the Analysis of the 4,977-bp and 7,436-bp Deletions and D-Loop Tandem Duplications in Human Mitochondrial DNA

Primer Pair	Amplified Position and Length	Length of PCR Product Amplified from mtDNA with a Length Mutation (bp)
L1-H1*	3304-3836 (533)	533
L2-H2**	8150-13650(5501)	524
L2-H3**	8150-13845(5696)	719
L4-H2**	8251-13650(5400)	423
L4-H4**	8251-14020(5770)	793
L4-H5•	8251-16228(7978)	542
L5-H5•	8344-16228(7885)	449
L4-H6•	8251-16430(8180)	744
L5-H6•	8344-16430(8087)	651
L6-H7Δ		200 or 260
L7-H8Δ		200 or 260

*: The primer set L1-H1 was used for the determination of the total mitochondrial DNA
**: The primer sets used for the determination of the 4,977-bp dmtDNA
•: The primer sets used for the determination of the 7,436-bp dmtDNA
Δ: The primer sets used for the determination of the tandem duplicated mtDNA:

L1 (3304–3323)	5′-AACATACCCATGGCCAACCT-3′
L2 (8150–8166)	5′-CCGGGGGTATACTACGGTCA-3′
L4 (8251–8270)	5′-GCCCGTATTTACCCTATAGC-3′
L5 (8344–8363)	5′-ACCAACACCTCTTTACAGTG-3′
L6 (336–355)	5′-AACACATCTCTGCCAAACCC-3′
L7 (467–486)	5′-CCCATACTACTAATCTCATC-3′
H1 (3836–3817)	5′-GGCAGGAGTAATCAGAGGTG-3′
H2 (13650–13631)	5′-GGGGAAGCGAGGTTGACCTG-3′
H3 (13845–13826)	5′-GTCTAGGGCTGTTAGAAGTC-3′
H4 (14020–14001)	5′-ATAGCTTTTCTAGTCAGGTT-3′
H5 (16228–16208)	5′-GTTGAGGGTTGATTGCTGTAC-3′
H6 (16430–16430)	5′-TGCGGGATATTGATTTCACG-3′
H7 (335–316)	5′-TAAGTGCTGTGGCCAGAAGC-3′
H8 (466–447)	5′-AGTGGGAGGGGAAAATAATG-3′

other investigators.[51,52] This indicates that our method of quantitation of the deleted mtDNA in human tissues is a reliable one.[19,51,52] Interestingly, the 4,977-bp deleted mtDNA started to appear in human testis and skin much later in one's life as compared to that seen in the brain, liver, or muscle.[19] These results clearly indicate that the 4,977-bp and 7,436-bp deleted mtDNAs are indeed accumulated in human tissues as a function of age and that there appears to be more deleted mtDNA accumulated in the tissues that turn over slowly or not at all (e.g., muscle) as compared to those that turn over relatively more rapidly (e.g., liver and blood cells). Moreover, we found that the tissues with

TABLE 3
Comparison of the Incidence of 4,977 bp and 7,436 bp dmtDNAs of Various Human Tissues

Age group (year)	4,977-bp dmtDNA (%) Muscle	Liver	Testis	7,436-bp dmtDNA (%) Muscle	Liver	Testis
0–9	0/4 (0.0)	N.A.	0/3 (0.0)	0/4 (0.0)	N.A.	0/3 (0.0)
10–29	4/7 (57.1)	1/2 (50.0)	0/5 (0.0)	1/7 (14.3)	1/2 (50.0)	0/5 (0.0)
30–39	4/6 (66.7)	4/5 (80.0)	0/6 (0.0	2/6 (33.3)	4/5 (80.0)	0/6 (0.0)
40–49	7/12 (58.3)	8/10 (80.0)	0/2 (0.0	4/12 (33.3)	9/10 (90.0	0/2 (0.0)
50–59	5/5 (100.0)	8/10 (80.0)	0/1 (0.0)	3/5 (60.0)	8/10 (80.0)	0/1 (0.0)
60–69	13/17 (76.5)	5/8 (62.5)	8/11 (72.7)	10/17 (58.8)	5/8 (62.5)	5/11 (45.5)
70–79	13/13 (100.0)	7/8 (87.5)	10/14 (71.4)	7/13 (53.8)	4/8 (50.0)	6/14 (42.9)
80–89	N.A.	N.A.	3/5 (60.0)	N.A.	N.A.	2/5 (40.0)

a high energy demand usually harbor higher proportions of the 4,977-bp and 7,436-bp deleted mtDNA as compared with the low-energy-demand tissues. It was also noted that the tissues with low mitotic activity bear higher level of the 4,977-bp deleted mtDNA. Moreover, the metabolic rate and the size of the tissue may be the other factors that affect the levels of the deleted mtDNA.[52]

AGING-ASSOCIATED TANDEM DUPLICATIONS IN THE D-LOOP OF mtDNA OF HUMAN TISSUES

As described previously, we detected, with PCR techniques, a 260-bp (type I) and a 200-bp (type II) tandem duplication in the mtDNA of muscle biopsies from aged individuals.[30] By use of several back-to-back primers, we further identified, with PCR and sequencing techniques, nine more types of tandem duplications in the D-loop region of mtDNAs from the brain, muscle, liver, and skin tissues of aged individuals.[44] These tandem duplications bear some interesting structural features. First, the start sites of the duplicated regions contain either a poly C run or a direct repeat sequence. Second, most of the insertion sites (where junctions occur) are localized at np 456, 494, and 568 where existent poly C runs in the D-loop. These observations have led us to propose that the tandem duplications are most likely generated by mispairing of the poly C runs or direct repeats near the replication origin O_H either at the beginning or at the end of replication of the heavy strand mtDNA. Thus, the duplicated regions are replicated one more time by the mitochondrial DNA polymerase.[44] The tandem duplications in the D-loop region may impair the function of mitochondria, because the D-loop is an important control region. It contains the replication origin of the heavy strand mtDNA and the promoters for RNA synthesis (LSP and HSP regulating the light and heavy strand transcription). The small tandem duplications may disturb the normal regulatory function of the control region for replication and transcription of mtDNA. This may as well perturb the synthesis and assembly of mtDNA-encoded polypeptides in the respiratory enzyme complexes, and thereby impair mitochondrial

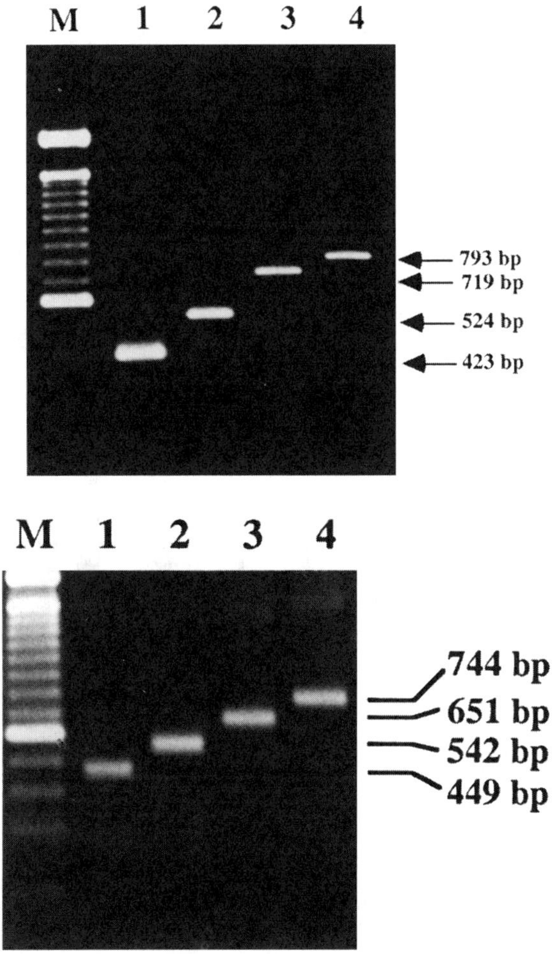

FIGURE 2. Electrophoretograms of the PCR products from human mtDNA with specific 4,977-bp deletion and 7,436-bp deletion. By using different primer pairs, the PCR products with different sizes were obtained and analyzed by electrophoresis on 1.5% agarose gel at 100 V for 45 min. DNA bands were then visualized under UV light transillumination after staining with ethidium bromide of the agarose gel. (A) Lane M, 100-bp DNA ladder; lane 1, 423-bp (L4-H2); lane 2, 524-bp (L2-H2); lane 3, 719-bp (L2-H3); and lane 4, 793-bp (L4-H4). (B) Lane M, 100-bp DNA ladder; lane 1, 449-bp (L5-H5); lane 2, 542-bp (L4-H5); lane 3, 651-bp (L5-H6); and lane 4, 744-bp (L4-H6).

respiration and ATP synthesis during human aging. About 90% of the subjects harboring the duplicated mtDNAs also had the 4,977-bp deletion.[30] Moreover, the incidences and proportions of the type II and type V duplications were found to be significantly increased with age.[30,44] The proportion of type II duplicated mtDNA in the muscle of a 71-year-old subject was 3.1% while that of a 55-year-old subject was only 0.78%. These two and several other types of

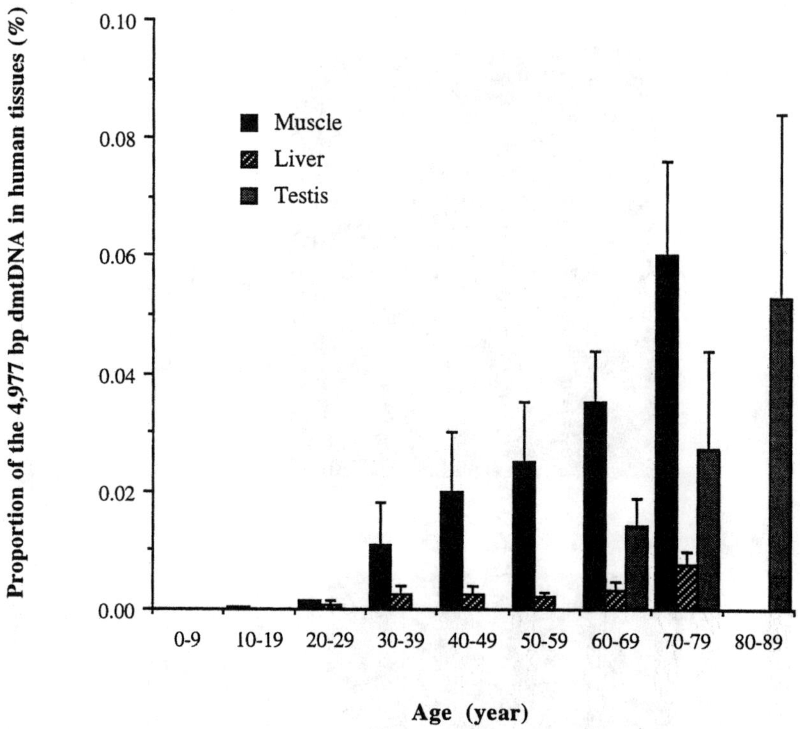

FIGURE 3. Age-dependent increase in the proportion of the 4,977-bp deleted mitochondrial DNA in human tissue. The black bar in each age group represents the average proportion (mean ± S.D.) of the 4,977-bp deleted mtDNA in the muscle tissues from subjects in the group. The PCR amplifications were all carried out under identical conditions. The number of subjects in muscle was 4, 5, 1, 6, 11, 4, 16, and 12 for the age group 0–9, 10–19, 20–29, 30–39, 40–49, 50–59, 60–69, and 70–79, respectively. The number of subjects in testis was 3, 4, 2, 6, 1, 1, 12, 12, and 6 for the age group 0–9, 10–19, 20–29, 30–39, 40–49, 50–59, 60–69, 70–79, and 80–89, respectively.

tandem duplications exist alone or in combination, and may cause synergistic deleterious effects on mitochondrial respiratory function.

OXIDATIVE DAMAGE AND HUMAN AGING

INTRACELLULAR SOURCE OF FREE RADICALS AND REACTIVE OXYGEN SPECIES

Free radicals have been described as molecular entities that contain at least one unpaired electron in a given atomic or molecular orbital. ROS such as superoxide anion, hydrogen peroxide, and hydroxyl radicals are formed *in vivo* during normal aerobic metabolism,[53,54] and may attack DNA, RNA, protein, and lipids. In turn, these reactive substances impair membrane structure and cell function. Accumulation of such damage may contribute to aging and age-associated degenerative disease.[6,10,21] The continuous threat of oxidative dam-

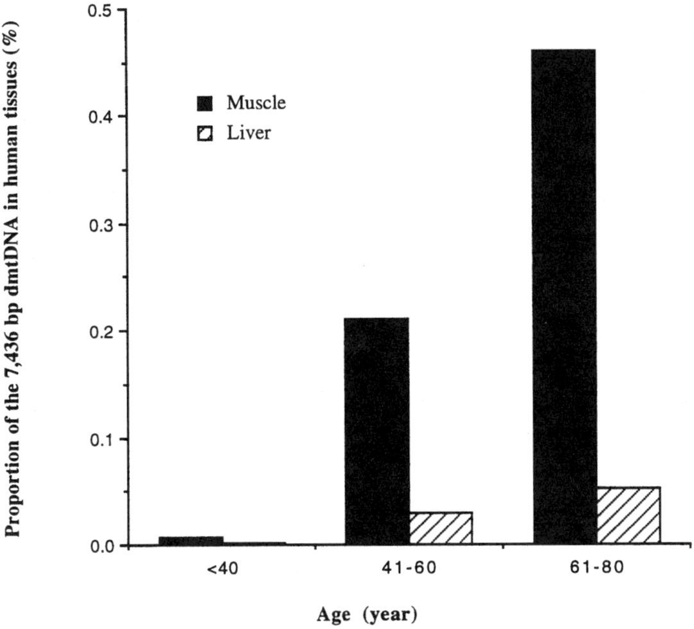

FIGURE 4. Age-dependent increase in the proportion of the 7,436-bp deleted mitochondrial DNA in human muscle and liver. The PCR amplifications were all carried out under identical conditions.

age to the cells, tissues, and the organism as a whole is underscored by the existence of an impressive array of cellular defenses that have evolved to combat these ROS.[5-7] However, these defense mechanisms are not perfect and, consequently, oxidative modifications of cellular biomolecules by escaped reactive oxygen species have been proposed to contribute significantly to aging.[1,6,7,21] Mitochondria are the most abundant source of ROS on the basis of the following evidence: (i) the mitochondrial electron transport system consumes approximately 90% of the oxygen utilized by human cells, and a relatively high steady-state level of incompletely reduced oxygen radicals is produced as a side-product in the matrix; (ii) in contrast to other oxidant-producing systems in human cells (e.g., cytochrome P450 system, various cytosolic oxidases, β-oxidation of fatty acids in peroxisomes), mitochondria are required for the production of ATP and are present in relatively high copy numbers in essentially all types of cells of the human body. Cellular energy deficits caused by the decline in mitochondrial respiratory function can impair normal cellular activities and compromise the cellular ability to cope with various challenges and stresses elicited by ROS. We argue that this oxidative damage to the mitochondria is one of the major factors in aging.

The cellular detoxification system for the ROS includes superoxide dismutase (SOD), catalase, glutathione peroxidase, and glutathione reductase. These enzymes comprise a major defense system against oxygen toxicity. The protective function of SOD appears to be restricted to a certain fraction of the total amount of the $O_2^{\bullet -}$-radicals produced.[21,53-55] Nohl and Hegner[53] demonstrated that increasing amounts of free radicals are accompanied by a concomitant increase of the lipoperoxide content of the biomembranes. This observation strongly supports the notion that lipid peroxidation may be promoted *in vivo,* and that the free-radical chain reaction exceeds the capacity of homeostatic protection as an animal or human ages. Recently, we investigated the content of mitochondrial lipoperoxides (the products of lipid peroxidation), which are damaging to many membrane-associated molecules with exposed sulfhydryl groups that may undergo intra- and inter-molecular cross-linking. Furthermore, we also analyzed the activities of the enzymes that catalyze the degradation of the harmful ROS such as manganese-superoxide dismutase (Mn-SOD) in the matrix of mitochondria. These studies have allowed us insights into the age-dependent changes in the intracellular levels of the ROS and the pertinent scavenging enzymes, which are very important in maintaining an appropriate redox status in the cell.[20]

AGE-DEPENDENT INCREASE IN THE OXIDATIVE DAMAGE OF DNA

Oxidative damage to DNA has been proposed to be a major cause of aging. Abundant evidence has been accumulated to support the idea that reactive oxygen species are an important causative factor in aging process: (i) ROS can react with specific DNA either at the sugar–phosphate backbone or at the nucleobase,[52-53] and cause single-strand DNA breaks; (ii) ROS induces chromosomal aberrations, and long-lived animal species have a slower rate of accumulation of these chromosomal aberrations; (iii) longer-lived animal species show a slower rate in the accumulation of lipofuscin pigments; and (iv) aging rate is proportional to the metabolic rate (cal/g/day) of the mammalian species.[7,55,56] Since the naked mtDNA is in proximity to the oxygen radicals generated as a by-product of mitochondrial respiration, it would be expected to be more susceptible to mutations than nuclear DNA.[19,29,57] MtDNA damage by ROS would elicit strand breakage with the possibility of intramolecular recombination or mis-pairing involving the direct repeats, as have been recently observed in our laboratory.[30,44] Such deletions and/or tandem duplications could impair mitochondrial respiratory function. It should also be noted that ischemia and other stresses may markedly increase the mtDNA deletions and the postmitotic, long-lived cells are most likely to accumulate the mutant mtDNA molecules.[52,58] On the other hand, numerous modifications of nucleobases have been detected when ROS reacts with DNA, which include 8-hydroxydeoxyguanosine (8OHdG), 5-hydroxyuracil, 5-hydroxycytosine, 4,6-diamino-5-formamidopyrimidine, 8-hydroxyadenine, thymine glycol, 2,6-diamino-5-formamido-pyrimidine, and 8-hydroxyadenine.[59,60] Among them, the most important and well-studied oxidized nucleobase is 8OHdG, which can be measured by several techniques in the femtomole range.[61] Besides being a useful marker for oxidative damage, the 8OHdG in DNA has been demonstrated to be associated with the potency of several mutagens.[57,62] It has been shown that 8OHdG is accumulated with age and, further, its accumulation has been suggested to be accelerated with the result of an age-associated accumulation of multiple mtDNA deletions.[63] The oxidatively modified mtDNA molecules cannot be efficiently repaired or removed and may be detrimental to cells that contain a high level of oxidatively modified DNA molecules.

AGE-DEPENDENT INCREASE OF LIPOPEROXIDE AND Mn-SOD ACTIVITY

Age-dependent increase in mitochondrial lipoperoxides

In recent years, we used the HPLC techniques to analyze the amount of lipoperoxides in submitochondrial particles (SMP) by the formation of the malonidialdehyde-thiobarbituric acid (MDA-TBA) adducts.[64] The average content of lipoperoxides in the SMP from tissues of the older subjects were found to be all significantly higher than those of young subjects. The relationship between age and lipoperoxide content of SMP from three different tissues are depicted in Figure 5. We also found that the average content of lipoperoxides in the SMP in the three tissues of the elderly subject groups were similar to one another. However, a higher level of lipoperoxides was observed in the muscle SMP of the subjects of the younger groups (<60 years old) as compared to those in the SMP of the liver and testes specimens from subjects of the corresponding age groups. This indicates that the lipoperoxide production is concurrently enhanced during aging in the muscle, liver, and testis (and probably the other tissues as well). The findings that a higher content of the MDA-TBA adduct exists in the muscle than in the liver and testis suggest that the amount of lipoperoxides and the extent of damage by free radicals on the membranes are greatest in the muscle when compared to liver and testes (Figure 5). This may be due to the fact that muscle is a tissue of high-energy demand and thus this tissue runs a high risk of being attacked by free radicals. Further, muscle cells turn over very slowly or not at all as compared to those that turn over more rapidly (e.g., liver, testis, and blood cells). Moreover, we found a positive correlation between the proportion of the 4,977-bp deleted mtDNA and MDA content in various human tissues (Figure 6).

Age-dependent increase in mitochondrial Mn-SOD activity

We determined the SOD activity on the basis of the extent of inhibition of the pyrogallol auto-oxidation by the tested sample.[65] The results showed that mitochondrial Mn-SOD activity is increased in an age-dependent manner.[20] We found that the average specific activities of Mn-SOD in the SMP from the tissues of the older subjects were significantly higher than those of the younger subjects (Figure 7). Moreover, we found that the average content of lipoperoxides in the SMP and the mitochondrial Mn-SOD activities in the liver, muscle, and

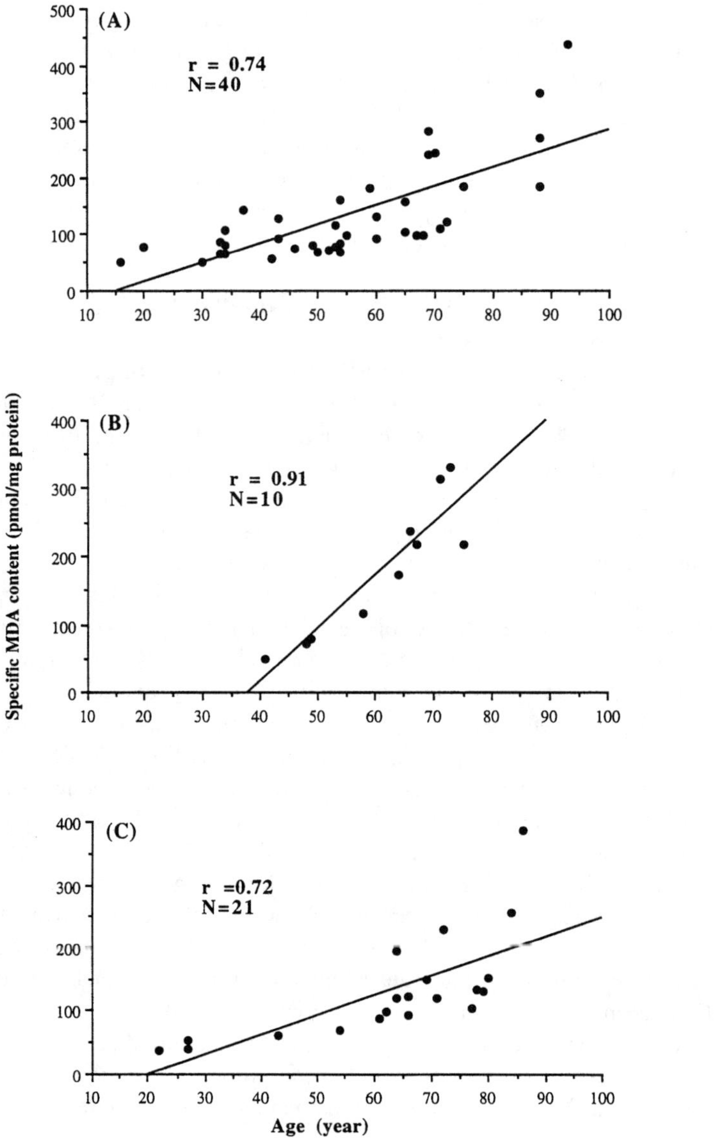

FIGURE 5. Correlation between the MDA content in the SMP and age of (A) muscle, (B) liver, and (C) testis tissues from human subjects of different ages.

testis of the subjects of the elderly groups were similar to one another (Figure 7). It is worth noting that the content of Mn-SOD of the liver is higher than that of the testis, and that the highest amount is in the high-energy-demand tissues such as the heart and kidney.[66] However, the detoxification of $O_2^{\bullet-}$ by SOD is not the terminal step; there are other factors and enzymes (e.g., catalase and

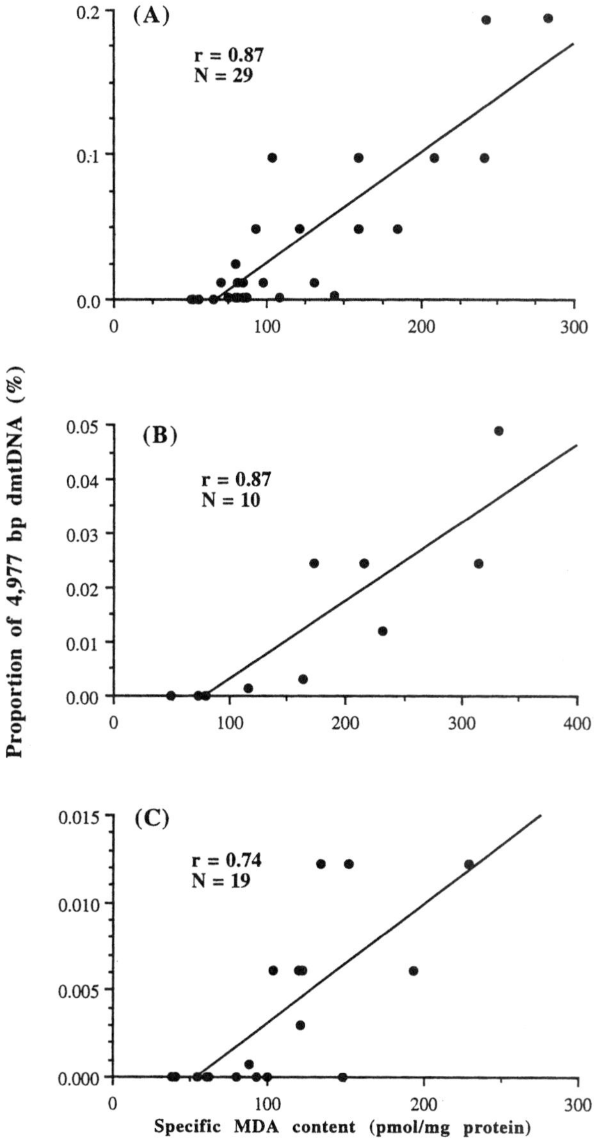

FIGURE 6. Correlation between the proportion of 4,977-bp deleted mtDNA and MDA content in the SMP of (A) muscle, (B) liver, and (C) testis tissues from human subjects of different ages.

glutathione peroxidase) involved in the detoxification process whose activities decline with the advanced age.[66] The presence of mitochondrial Mn-SOD enables the cell to tolerate a small but significant fraction of the oxygen in the mitochondria being reduced by the one-electron transfer process. The arrangement of the electron transport system in the mitochondrial inner membranes

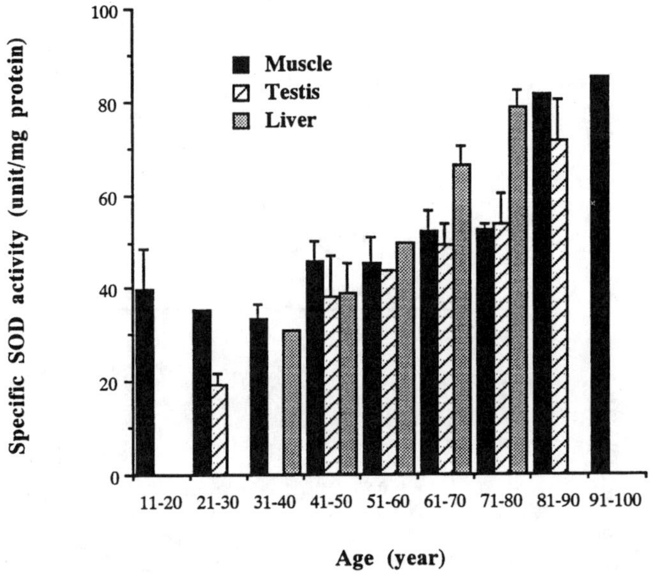

FIGURE 7. Age-dependent increase of the manganese superoxide dismutase (Mn-SOD) in the muscle, liver, and testis of the human. The bars in each age group represents the average Mn-SOD specific activity in the three tissues. All of the values are expressed as mean ± SEM. The Mn-SOD activity was measured as described by Yen et al. (Ref. 20). The number of subjects for the muscle was 2, 1, 6, 4, 7, 7, 4, 1, and 1 for the age groups 11–20, 21–30, 31–40, 41–50, 51–60, 61–70, 71–80, 81–90, and 91–100, respectively. The number of subjects for the testis was 3, 2, 1, 7, 6, and 3 for the age groups 21–30, 41–50, 51–60, 61–70, 71–80, and 81–90, respectively. The number of subjects for the liver was 1, 4, 1, 4, and 3 for the age groups 31–40, 41–50, 51–60, 61–70, and 71–80, respectively.

favors the concept that this is indeed the system that accounts for the generation of most, if not all, of the ROS that are responsible for peroxidation of membrane lipids *in vivo*.[53,67]

DISCUSSION

Aging is the late phase of the natural process of cellular differentiation and normal development of the human body that culminates in death. Some biomedical scientists believe that aging is the manifestations of a progressive loss of physical and mental acuity caused by the impairment of physiological functions by ill-defined factors.[68] In 1956, Harman[2] proposed that free radicals are likely involved in the aging process. Over the past four decades, strong evidence has been accumulated to support the notion that free radicals cause many of the functional alterations seen in human aging and in humans with a variety of degenerative diseases.

Free radicals randomly attack various cellular constituents and thereby impair cell function. They damage the membrane integrity and result in an accumulation of oxidized dysfunctional proteins.[68] Free radicals attack DNA and result in an increase of DNA damage.[63,69] In the past decade, it has been shown that aging is associated with an enhanced production *in vivo* of modified purine and pyrimidine bases,[27] which are presumably degradative products from the excision and repair of damaged nucleic acids. Recent studies have suggested that accumulation of mtDNA mutations in the heart, and probably other tissues as well, most likely results from oxidative damage caused by superoxide anions and hydroxyl radicals.[21,70] Since the naked and fast-replicating mtDNA is located in the mitochondrial matrix close to the origins of free radicals, it has been shown to be much more vulnerable to oxidative damage than nuclear DNA.[6,7,19,21]

Recently, we have developed several methods to identify the existence of mtDNA deletions and to determine the proportions of 4,977-bp and 7,436-bp deleted mtDNAs. The results obtained so far indicate that the frequency of occurrence and proportions of the 4,977-bp and 7,436-bp deleted mtDNAs are higher in the muscle than those in the liver and testis in human aging.[19] Moreover, we found that there is a positive correlation between lipoperoxide content and proportion of 4,977-bp deleted mtDNA in various tissues of the aging human (Figure 6). This finding is easily rationalized because defective mitochondrial respiratory chain, caused by mtDNA mutations, will generate more ROS that will in turn damage the phospholipids of the mitochondrial inner membranes. Thus, a vicious cycle of electron leak, mtDNA mutation, and enhanced lipid peroxidation will be operating continuously during aging in essentially all the tissues. Thus, the 4,977-bp and 7,436-bp deleted mtDNAs may serve as specific convenient genetic markers (scars) to monitor the molecular changes in human tissues during the aging process. Even more interesting, we have recently demonstrated that the frequency of occurrence and proportion of the 4,977-bp deleted mtDNA are well correlated with the decline of motility and fertility of human sperm.[71] The sperm with poor motility had a significantly higher amount of the 4,977-bp deleted mtDNA and lipid peroxides measured as MDA-TBA adduct. It is worth mentioning that the Mn-SOD activity in the mitochondria showed an age-dependent increase (Figure 7), which may be a safeguard response of tissue cells to the ever-increasing amounts of lipoperoxides in the mitochondria during human aging.[20] Determination of the modified bases in mitochondria and the changes in the activities of the detoxification enzyme system including superoxide dismutase, glutathione peroxidase, glutathione reductase, and catalase in human tissues from subjects of different ages will provide further information in understanding the detailed underlying mechanism of the action of ROS and the role of mitochondrial electron leak and mtDNA mutations in the aging process of the human.

CONCLUSIONS

Mitochondrial DNA is a naked double-strand circular extrachromosomal genetic element continuously exposed in a matrix that contains large amounts of ROS and free radicals. Mitochondria are the most important intracellular source of ROS and free radicals (e.g., ubisemiqinone radicals), and mitochondria DNA is heavily damaged by ROS at the nucleobases, as indicated by the high steady-state level of 8-OHdG, the presence of which causes mispairing and point mutation in mtDNA. The age-dependent decline in the capability and capacity of mitochondria to dispose of the noxious ROS will render mtDNA more vulnerable to various types of mutations and oxidative modifications during aging. During the past five years, more than two dozen different types of large-scale deletions in mtDNA of various tissues of aged individuals have been identified by molecular biological techniques. The 4,977-bp and 7,436-bp deletions are the most prevalent and abundant ones among these deletions. The most common one is the 4,977-bp deletion with a 13-bp direct repeat flanking the 5′- and 3′-end breakpoints at np 8470/8482 and np 13447/13459, respectively. The onset age of the various deletions in mtDNA varies greatly with individuals and the type of the deletions. By using the PCR techniques, the frequency of occurrence and the proportion of the 4,977-bp and 7,436-bp deleted mtDNAs were found to increase with age. Several investigators have proposed that accumulation of mutations in mtDNA is the major cause of respiratory function decline during the aging process. Recently, ten heteroplasmic tandem duplications in the D-loop region of mtDNAs have been identified in our laboratory in aging human tissues by PCR techniques using back-to-back primers. These tandem duplications, occurring alone or in combination, may perturb the replication and/or transcription of mtDNA and in turn impair the respiratory function of mitochondria in human aging. Similar age-associated length mutations of mtDNAs have also been observed in aged animals and lower organisms. On the other hand, we found that the level of lipoperoxides and Mn-SOD activity in the mitochondria exhibited an age-dependent increase in various human tissues. Furthermore, there exists a positive correlation between the proportion of the mutant mtDNAs and the level of lipoperoxides in the mitochondria. Given these results and pertinent data reported from several other laboratories together, we suggest that enhanced generation of ROS and lipid peroxides in the mitochondria during the aging process may cause oxidative damage and large-scale deletions or insertions of mtDNA, and that the broad spectrum of mtDNA mutations or modifications are early molecular events and major contributions to aging in humans.

REFERENCES

1. **Harman, D.,** The aging process: major risk factor for the disease and death, *Proc. Natl. Acad. Sci. U.S.A.,* 88, 5360, 1991.
2. **Harman, D.,** Aging: theory based on free radical and radiation chemistry, *J. Gerontol.,* 11, 298, 1956.
3. **Burkitt, M. J. and Mason, P. R.,** Direct evidence for in vivo hydroxy-radical generation in experimental iron overload: an ESR spin-trapping investigation, *Proc. Natl. Acad. Sci. U.S.A.,* 88, 8440, 1991.

4. **Ames, B. N.,** Endogenous DNA damage as related to cancer and aging, *Mutat. Res.,* 214, 41, 1989.
5. **Gutteridge, J. M. C., Westermarck, T., and Halliwell, B.,** Oxygen radical damage in biological systems, in *Free Radicals, Aging, and Degenerative Disease,* Johnson, Jr., J. E., Walford, R., Harman, D., and Miquel, J., Eds., Alan R. Liss, New York, 1985, 99.
6. **Sual, R. L. and Ames, B. N.,** Endogenous DNA damage as related to cancer and aging, *Mutat. Res.,* 214, 41, 1989.
7. **Aldeman, R., Saul, R. L., and Ames, B. N.,** Oxidative damage to DNA: relation of species metabolic rate and life span, *Proc. Natl. Acad. Sci. U.S.A.,* 85, 2706, 1988.
8. **Comporti, M.,** Biology of disease: lipid peroxidation and cellular damage in toxic liver injury, *Lab. Invest.,* 53, 599, 1985.
9. **Luft, R.,** The development of mitochondrial medicine, *Proc. Natl. Acad. Sci. U.S.A.,* 91, 8731, 1994.
10. **Wei, Y. H.,** Mitochondrial DNA alterations as aging-associated molecular events, *Mutat. Res.,* 275, 145, 1992.
11. **Zhang, C., Baumer, A., Maxwell, R. J., Linnane, A. W., and Nagley, P.,** Multiple mitochondrial DNA deletions in an elderly human individual, *FEBS Lett.,* 297, 34, 1992.
12. **Ikebe, S., Tanaka, M., Ohno, K., Sato, W., Hattori, K., Kondo, T., Mizuno, Y., and Ozawa, T.,** Increase of deleted mitochondrial DNA in the striatum in Parkinson's disease and senescence, *Biochem. Biophys. Res. Commun.,* 170, 1044, 1990.
13. **Yen, T. C., Su, J. H., King, K. L., and Wei, Y. H.,** Aging-associated 5 kb deletion in human liver mitochondrial DNA, *Biochem. Biophys. Res. Commun.,* 178, 124, 1991.
14. **Yen, T. C., Pang, C. Y., Hsieh, R. H., Su, C. H., King, K. L., and Wei, Y. H.,** Age-dependent 6 kb deletion in human liver mitochondrial DNA, *Biochem. Intl.,* 26, 457, 1992.
15. **Wallace, D. C.,** Mitochondrial DNA sequence variation in human evolution and disease, *Proc. Natl. Acad. Sci. U.S.A.,* 91, 8739, 1994.
16. **Zhang, C., Linnane, A. W., and Nagley, P.,** Occurrence of a particular base substitution (3243 A to G) in mitochondrial DNA of tissues of aging humans, *Biochem. Biophys. Res. Commun.,* 195, 1104, 1993.
17. **Munscher, C., Rieger, T., Muller-Hocker, J., and Kadenbach, B.,** The point mutation of mitochondrial DNA characteristic for MERRF disease is found also in healthy people of different ages, *FEBS Lett.,* 317, 27, 1993.
18. **Yen, T. C., Chen, Y. S., King, K. L., Yeh, S. H., and Wei, Y. H.,** Liver mitochondrial respiratory functions decline with age, *Biochem. Biophys. Res. Commun.,* 165, 994, 1989.
19. **Lee, H. C., Pang, C. Y., Hsu, H. S., and Wei, Y. H.,** Differential accumulations of 4,977 bp deletion in mitochondrial DNA of various tissues in human aging, *Biochim. Biophys. Acta,* 1226, 37, 1994.
20. **Yen, T. C., King, K. L., Lee, H. C., Yeh, S. H., and Wei, Y. H.,** Age-dependent increase of mitochondrial DNA deletions together with lipid peroxides and superoxide dismutase in human liver mitochondria, *Free Rad. Biol. Med.,* 16, 207, 1994.
21. **Shigenaga, M. K., Hagen, T. M., and Ames, B. N.,** Oxidative damage and mitochondrial decay in aging, *Proc. Natl. Acad. Sci. U.S.A.,* 91, 10771, 1994.
22. **Benzi, G., Pastoris, O., Marzatico, F., Villa, R. F., Dagani, F., and Curti, D.,** The mitochondrial electron transfer alteration as a factor involved in the brain aging, *Neurobiol. Aging,* 13, 361, 1992.
23. **Trounce, I., Byrne, E., and Marzuki, S.,** Decline in skeletal muscle mitochondrial respiratory chain function with aging, *Lancet* i, 673, 1989.
24. **Attardi, G.,** Animal mitochondrial DNA: an extreme example of genetic economy, *Int. Rev. Cytol.,* 93, 93, 1984.
25. **Monnat, R. J. and Loeb, L. A.,** Nucleotide sequence preservation of human mitochondrial DNA, *Proc. Natl. Acad. Sci. U.S.A.,* 82, 2895, 1985.
26. **Linnane, A. W., Marzuki, S., Ozawa, T., and Tanaka, M.,** Mitochondrial DNA mutations as an important contributor to aging and degenerative diseases, *Lancet* i, 642, 1989.
27. **Bandy, B. and Davison, A. J.,** Mitochondrial mutations may increase oxidative stress: implications for carcinogenesis and aging? *Free Rad. Biol. Med.,* 8, 523, 1990.

28. **Fukunaga, M. and Yielding, K. L.,** Fate during cell growth of yeast mitochondrial and nuclear DNA after photolytic attachment of the monoazide analog of ethidium bromide, *Biochem. Biophys. Res. Commun.,* 90, 582, 1979.
29. **Wallace, D. C., Ye, J. H., Neckelmann, S. N., Singh, G., Webster, K. A., and Greenberg, B. D.,** Sequence analysis of cDNAs for the human and bovine ATP synthase b-subunit: mitochondrial genes sustain seventeen times more mutations, *Curr. Genet.,* 12, 81, 1987.
30. **Lee, H. C., Pang, C. Y., Hsu, H. S., and Wei, Y. H.,** Aging-associated tandem duplication in the D-loop of mitochondrial DNA of human muscle, *FEBS Lett.,* 354, 79, 1994.
31. **Ferrandiz, M. L., Martinez, M., Juan, E. D. Diez, A., Bustos, G., and Miquel, J.,** Impairment of mitochondrial oxidative phosphorylation in the brain of aged mice, *Brain Res.,* 644, 335, 1994.
32. **Holt, I. J., Harding, A. E., and Morgan-Hughes, J. A,** Deletions of mitochondrial DNA in patients with mitochondrial myopathies, *Nature,* 331, 771, 1988.
33. **Larsson, N. G., Holme, E., Kristansson, B., Oldfords, A., and Tulinius, M.,** Progressive increase of mutated mitochondrial DNA fraction in Kearns-Sayre syndrome, *Pediatr. Res.,* 28, 131, 1990.
34. **Holt, I. J., Harding, A. E., Cooper, J. M., Schapira, A. H. V., Toscano, A., Clark, J. B., and Morgan-Hughes, J. A.,** Mitochondrial myopathies: clinical and biochemical features of 30 patients with major deletions of muscle mitochondrial DNA, *Ann. Neurol.,* 29, 699, 1989.
35. **Ohno, K., Tanaka, M., Sahashi, K., Ini, T., Sato, W., Tamamoto, T., Takahashi, A., and Ozawa, T.,** Mitochondrial DNA deletions in inherited recurrent myoglobinuria, *Ann. Neurol.,* 29, 364, 1991.
36. **Schon, E. A., Rizzuto, R. M., Moraes, C. T., Nakase, H., Zeviani, M., and Dimauro, S.,** A direct repeat is a hotspot for large-scale deletion of human mitochondrial DNA, *Science,* 244, 346, 1988.
37. **Blok, R. B., Thorburn, D. R., Thompson, G. N., and Dahl, H. H.,** A topoisomerase II cleavage site is associated with a novel mitochondrial DNA deletion, *Hum. Genet.,* 95, 75, 1995.
38. **Poulton, J., Deadman, M. E., and Gardiner, R. M.,** Duplication of mitochondrial DNA in mitochondrial myopathy, *Lancet* i, 236, 1989.
39. **Rotig, S. A., Cormier, V., Blanche, S., Bonnefont, J. P., and LeDeist, F.,** Pearson's marrow-pancreas syndrome: a multisystem mitochondrial disorder in infancy, *J. Clin. Invest.,* 86, 1601, 1990.
40. **Rotig, S. A., Bessis, J. L., Romero, N., Cormier, V., Saudubray, J. M., Narcy, P., Lenoir, G., Rustin, P., and Munnich, A.,** Maternally inherited duplication of mitochondrial genome in a syndrome of proximal tubulopathy, diabetes mellitus, and cerebral ataxia, *Am. J. Hum. Genet.,* 50, 364, 1992.
41. **Poulton, J. and Gardiner, R. M.,** Non-invasive diagnosis of mitochondrial myopathy, *Lancet* i, 961, 1989.
42. **Brockington, M., Sweeney, M. G., Hammans, S. R., Hughes, J. A., and Harding, A. E.,** A tandem duplication in the D-loop of human mitochondrial DNA is associated with deletions in mitochondrial myopathies, *Nat. Genet.,* 4, 67, 1993.
43. **Poulton, J. and Holt, I. J.,** Mitochondrial DNA: does more lead to less? *Nat. Genet.,* 8, 313, 1995.
44. **You, B. J.,** Studies on the tandem duplications in the D-loop region of human mitochondrial DNA, Master's thesis, Institute of Biochemistry, National Yang-Ming University, Taipei, 1995.
45. **Wallace, D. C., Singh, G., Lott, M. T., Hodge, J. A., Schurr, T. G., Lezza, A. M. S., Elsas, L. J. II., and Nikoskelainen, E. K.,** Mitochondrial DNA mutation associated with Leber's Hereditary Optic Neuropathy, *Science,* 242, 1427, 1988.

46. **Hess, J. F., Parisis, M., Bennet, J. L., and Clayton, D. A.,** Impairment of mitochondrial transcription termination by a point mutation associated with the MELAS subgroup of mitochondrial encephalopathies, *Nature,* 351, 236, 1991.
47. **Shoffner, J. M., Lott, M., Hodge, J., Lezza, A. M. S., Seibel, P., Ballinger, S. W., and Wallace, D. C.,** Myoclonic Epilepsy and ragged-red fibre disease (MERRF) is associated with mitochondrial DNA $tRNA^{Lys}$ mutation, *Cell,* 61, 931, 1990.
48. **Holt, I. J., Harding, A. E., Petty, R. K., Morgan-Hughes, J. A.,** A new mitochondrial disease associated with mitochondrial DNA heteroplasmy, *Am. J. Hum. Genet.,* 46, 428, 1990.
49. **Tatuch, Y., Christodoulou, J., Feigenbaum, A., Clarke, J. T. R., Wherret, J., Smith, C., Rudd, N., Petrovabenedict, R., and Robinson, B. H.,** Heteroplasmic mtDNA mutation (T Æ G) can cause Leigh disease when the percentage of abnormal mtDNA is high, *Am. J. Hum. Genet.,* 50, 852, 1992.
50. **Torroni, A. and Wallace, D. C.,** Mitochondrial DNA variation in human populations and implications for detection of mtDNA mutations of pathological significance, *J. Bioenerg. Biomembr.,* 26, 261, 1994.
51. **Ozawa, T., Tanaka, M., Ikebe, S., Hono, K., Kondo, T., and Mizuno, Y.,** Quantitative determination of deleted mitochondrial DNA relative to normal DNA in Parkinson striatum by a kinetic PCR analysis, *Biochem. Biophys. Res. Commun.,* 172, 483, 1990.
52. **Cortopassi, G. A., Shibata, D., Soong, N. W., and Arnheim, N.,** A pattern of accumulation of a somatic deletion of mitochondrial DNA in aging human tissues, *Proc. Natl. Acad. Sci. U.S.A.,* 89, 7370, 1992.
53. **Nohl, H. and Hegner, D.,** Do mitochondria produce oxygen radicals in vivo? *Eur. J. Biochem.,* 82, 563, 1978.
54. **Chance, B., Sies, H., and Boveris, A.,** Hydroperoxide metabolism in mammalian organs, *Physiol. Rev.,* 59, 527, 1979.
55. **Culter, R. G.,** Human longevity and aging: possible role of reactive oxygen species, *N.Y. Acad. Sci.,* 621, 1, 1991.
56. **Imlay, J. A. and Linn, S.,** DNA damage and oxygen radical toxicity, *Science,* 240, 1302, 1988.
57. **Shibutani, S., Takeshita, M., and Grollman, A. P.,** Insertion of specific bases during DNA synthesis past the oxidation damaged bases 8-oxo dG, *Nature,* 349, 431, 1991.
58. **Marin, G. R., Danner, D. B., and Holbrook, N. J.,** Aging-causes and defenses, *Ann. Rev. Med.,* 44, 419, 1993.
59. **Dizdaroglu, M.,** Oxidative damage to DNA in mammalian chromatin, *Mutat. Res.,* 275, 331, 1992.
60. **Jackson, J. H., Gajewski, E., Schraufstatter, I. U., Hyslop, P. A., Fuciarelli, A. F., Cochrane, C. G., and Dizdaroglu, M.,** Damage to the bases in DNA induced by stimulated human neutrophils, *J. Clin. Invest.,* 84, 1644, 1989.
61. **Halliwell, B. and Aruoma, O. I.,** DNA damage by oxygen-derived species. Its mechanism and measurement in mammalian systems, *FEBS Lett.,* 281, 9, 1991.
62. **Cheng, K. C., Cahill, D. S., Kasaim H., Nishimura, S., and Loeb, L.,** 8-Hydroxyguanine, an abundant form of oxidative DNA damage, causes GT and AC substitution, *J. Biol. Chem.,* 267, 166, 1992.
63. **Hayakawa, M., Torri, K., Sugiyama, S., Tanaka, M., and Ozawa, T.,** Age-associated accumulation of 8-hydroxydeoxyguanosine in mitochondrial DNA of human diaphragm, *Biochem. Biophys. Res. Commun.,* 180, 894, 1991.
64. **Wong, S. H. Y., Knight, J. A., Hopfer, S. M., Zaharia, O., Leach, C. N., and Sunderman, F. W.,** Lipoperoxides in plasma as measured by liquid-chromatographic separation of malondialdehyde-thiobarbituric acid adduct, *Clin. Chem.,* 33, 214, 1987.
65. **Violi, F. I., Alessandri, C., and Ghiselli, A.,** A simple method for evaluating platelet superoxide dismutase, *Scand. J. Clin. Lab. Invest.,* 45, 351, 1985.

66. **Taniguchi, N.,** Clinical significances of superoxide dismutases: changes in aging, diabetes, ischemia, and cancer, *Adv. Clin. Chem.,* 29, 1, 1992.
67. **Richter, C., Park, J. W., and Ames, B. N.,** Normal oxidative damage to mitochondrial and nuclear DNA is extensive, *Proc. Natl. Acad. Sci. U.S.A.,* 85, 6465, 1988.
68. **Stadtman, E. R.,** Protein oxidation and aging, *Science,* 257, 1220, 1992.
69. **Higuchi, Y. and Linn, S.,** Purification of all forms of HeLa cell mitochondrial DNA and assessment of damage to it caused by hydrogen peroxide treatment of mitochondria or cells, *J. Biol. Chem.,* 270, 7950, 1995.
70. **Miquel, J.,** An integrated theory of aging as the result of mitochondrial DNA mutation in differentiated cells, *Arch. Gerontol. Geriatr.,* 12, 99, 1991.
71. **Kao, S. H., Chao, H. T., and Wei, Y. H.,** Mitochondrial deoxyribonucleic acid 4,977-bp deletion is associated with diminished fertility and motility of human sperm, *Biol. Reprod.,* 52, 729, 1995.
72. **Hsieh, R. H., Hou, J. H., Hsu, H. S., and Wei, Y. H.,** Age-dependent respiratory function decline and DNA deletions in human muscle mitochondria, *Biochem. Mol. Biol. Intl.,* 32, 1009, 1994.

Chapter 10

REGULATION OF APOPTOTIC EVENTS BY ZINC

Pamela J. Fraker and William Telford

CONTENTS

GENERAL CHARACTERISTICS OF APOPTOSIS OR PROGRAMMED CELL DEATH

Apoptosis is a programmed form of cell death or suicide which plays a key role in the development and regulation of many tissues (see Table 1). In the last decade, it has also become apparent that apoptosis modulates many facets of immune function.[1] For example, suboptimal levels of cytokines or exposure to ionizing radiation, glucocorticoids, various drugs, etc., can induce 30 to 70% of immature thymocytes and developing cells of the B-lineage in the marrow of mice to undergo apoptosis.[1,2] Thus, immature lymphocytes seem very vulnerable to apoptotic cues. Indeed, in the last few years, glucocorticoid-induced

0-8493-9485-6/96/$0.00+$.50

TABLE 1
Biological and Immunological Roles Played by Apoptosis

I. Immune Processes Modulated by Apoptosis
 Elimination of anti-self clones
 Removal of developing T- and B-cells with faculty gene rearrangements
 Glucocorticoid-mediated death of thymocytes
 Glucocorticoid-mediated death of cells of the B-lineage of the bone marrow
 Irradiation-induced death of thymocytes and B-lineage cells of the marrow
 Destruction of cancer cells by natural killer and cytolytic T-cells
 Elimination of B-cells by antibodies that cross-link surface IgM molecules
 Elimination of T-cells by antibodies which bind to CD3 or the T-cell receptor

II. Developmental Processes Regulated by Apoptosis
 Elimination of rudimentary tissues
 Removal of excess neurons
 Metamorphosis
 Turnover of skin cells
 Shedding of uterine lining
 Castration

III. Chemotherapy Agents That Induce Apoptosis
 Cis-platin
 Etoposide
 Camptothecin
 Methotrexate
 5-Fluorouracil

IV. Diseases or Disease States Where Apoptosis Plays a Role
 Cancer
 AIDS
 Malnutrition

death of thymocytes has become a model system frequently used to study apoptosis. This form of cell death is also important to the removal of dangerous anti-self clones and B-cells which exhibit faulty rearrangement of immunoglobulin genes.[1] Of particular importance is the discovery that natural killer and cytolytic T-cells destroy malignant cells by inducing them to undergo apoptosis.[1] Indeed, oncologists are finding that some of the more efficacious chemotherapeutic drugs act by inducing high levels of apoptosis in malignant cells.[3] Conversely, hard to treat cancers appear to acquire an array of oncogenes or protooncogenes that make them less prone to respond to death cues and more likely to remain immortal.[4] It now appears that apoptosis plays a role in the etiology of AIDS by promoting heightened levels of cell death in helper T-cells.[5] More recently, our laboratory has shown that the chronically elevated glucocorticoids generated during malnutrition, especially deficiencies in protein-calories or zinc, are sufficient to induce extensive apoptosis in developing thymocytes and cells of the B-lineage in the marrow.[2-6] This may account for the thymic atrophy and lymphopenia frequently noted in malnourished humans and animals. In addition, *in vitro* deprivation of zinc caused apoptosis to occur in thymocytes and myeloid cell lines.[7,8] Thus, apoptosis may play a major role in the depletion of cells of the immune system during malnutrition.[9] During development, rudimentary tissues and excess neurons are eliminated via apoptotic mechanisms.

Cells which die necrotically or accidentally usually enlarge in size with considerable swelling noted among internal organelles. Interestingly, cells which die apoptotically condense in size. The nuclei of cells destined to die begin retracting with pronounced changes noted in the distribution of the chromatin. Nuclear lamina disassembly accompanies these physical changes.[10] This is quickly followed by induction of an endogenous endonuclease that digests the DNA of the cell at the linker regions near the nucleosomes, generating small 180-bp multimers of DNA.[11] These multiple small fragments of DNA or the so-called "ladders" of DNA, have became the primary hallmark used by many cell biologists to verify the occurrence of apoptotic death.[1,11]

Extraction and gel electrophoresis of DNA to verify the existence of DNA ladders or small fragments of DNA takes three to four days and does not provide quantitative information with regard to the proportion of cells in the population that are, in fact, undergoing apoptosis. Conversely, multiparameter flow cytometry provides a rapid, highly quantitative means of determining the precise proportion of cells in a population that are undergoing apoptosis in less than three hours (Figure 1). Using cell cycle analysis, our lab demonstrated that apoptotic cells appeared in the so-called hypodiploid region of the cell cycle due to the fact that they exhibit less DNA fluorescence.[6] A simple example is shown in Figure 1, where most normal thymocytes are in the G_0/G_1 phase of the cell cycle. Apoptotic cells induced by treating the thymocytes with a glucocorticoid fall into a distinct new region. We demonstrated that cells in this region, now called A_0, had fragmented DNA and morphological characteristics of apoptotic cells.[12] Moreover, by using fluorescently-labeled antibodies directed at specific cell surface markers for a specific subpopulation of cells of interest, flow cytometry now makes it possible to measure apoptosis in a subset of cells within a heterogenous tissue.[2,12] For example, by using fluorescently-labeled antibody against IgM, one can now monitor glucocorticoid-induced apoptosis among immature and mature B-cells in the bone marrow.[2] The limitations of past methodology made such studies impossible since these cells represent less than 10% of the marrow. Further information on this methodology and the many systems where flow cytometry has been employed for the study of apoptosis are reviewed elsewhere.[12] As a word of caution, a few examples of apoptosis do exist where the dying cells exhibited the characteristic morphology of an apoptotic cell though no DNA ladder was generated.[13] Thus, DNA fragmentation is not a completely reliable indicator for apoptotic death.

Conversely, pathologists have focused on the blebbing of the membrane of the cell and the generation of the characteristic apoptotic bodies, which morphologically defines this form of cell death.[1] Changes in the cell membrane of the dying cell can include heightened expression on the cell's surface of

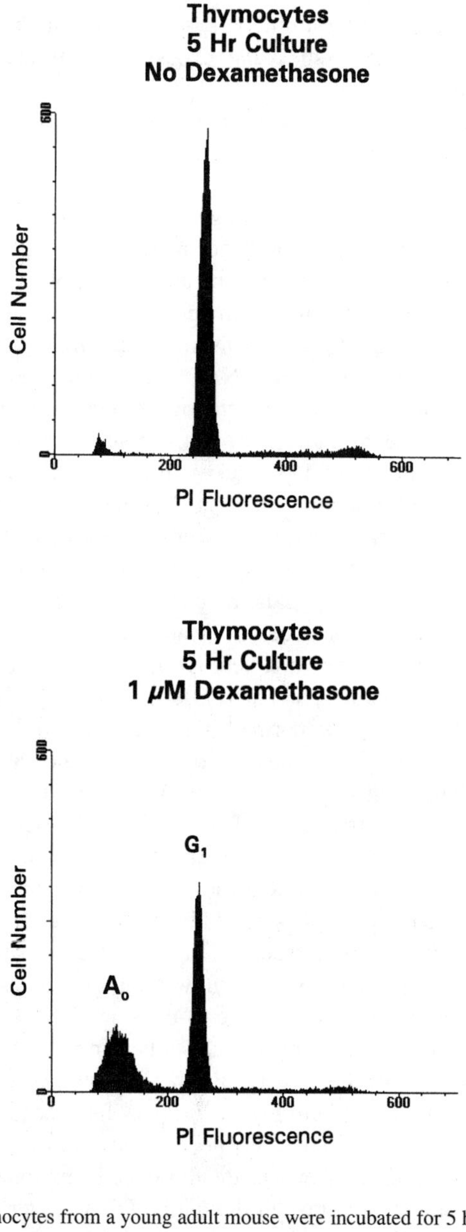

FIGURE 1. Thymocytes from a young adult mouse were incubated for 5 hr with and without 1 m*M* dexamethasone, a glucocorticoid commonly used to induce apoptosis. Subsequently, the cells were washed and fixed with 70% ethanol. Prior to cell cycle analysis on the flow cytometer, the DNA of the cell was stained with propidium iodide (PI). Since thymocytes do not cycle much, most of the untreated cells in the top panel fell into the G_o/G_1 region. Due to extensive fragmentation of their DNA, apoptotic cells exhibited less DNA fluorescence, falling into a hypodiploid region called A_o to the left the G_o/G_1 peak. In the case of untreated thymocytes, only a small portion of the population was undergoing cell death (4%), whereas in the bottom panel dexamethasone had induced 41% of the thymocytes to enter the apoptotic state.

phosphatidylserine and some glycans. These changes quickly invite the attention of phagocytic cells that rapidly remove the dying cells.[1,11] Activation of transglutaminases causes extensive cross linkage of proteins in the cell, which may contribute to the altered morphology of the apoptotic cell and perhaps prevent leakage of large amounts of intracellular material. Classical apoptosis also appears to be programmed and dependent on the *de novo* expression of so-called death genes. Thus, inhibitors of RNA or protein synthesis such as actinomycin D or cycloheximide are commonly used to block apoptosis.[11] As will be discussed, high concentrations of zinc can also block apoptotic death in a variety of systems.[1,6,11] Most of the organelles of the apoptotic cell appear intact and are presumably operable, at least, in the early phase of death.[11] Clearly, the central role that apoptosis plays in development and function of the immune system, as well as various diseases, indicates that it is an important phenomenon of keen interest to life scientists. We will provide evidence herein that apoptosis is a phenomenon that should be of particular interest to nutritionists and trace element biochemists.

ANTAGONISTIC EFFECTS OF HIGH CONCENTRATIONS OF ZINC ON APOPTOSIS

The literature shows an interesting interplay between zinc and apoptosis. Over a decade ago, it was discovered that high levels of zinc (500 to 1000 μ*M)* would block apoptosis in thymic nuclei. It was postulated that zinc inhibited the endogenous endonuclease.[14] Subsequently, high, nonphysiological concentrations of zinc (500 μ*M)* have been found to inhibit apoptotic death induced by a wide variety of stimuli in a variety of cell types. In the case of murine thymocytes, apoptotic stimuli as diverse as glucocorticoids, gamma radiation, adenosine, and ATP have been inhibited by high zinc.[14-17] High zinc also inhibited glucocorticoid- and gamma-radiation-induced apoptosis in mouse bone marrow B-lineage lymphocytes, demonstrating that it has a significant regulatory effect on apoptotic events in the progenitor cells of the immune system.[2] Zinc at 1 m*M* inhibited TNF-induced death in WEHI 164S fibrosarcoma cell line and at 500 to 1 μ*M* also inhibited sporodesmin and gliotoxin-induced cell death in T-lymphoblasts and macrophages, respectively.[18,19] Etoposide- and UV radiation-induced cell death in HL-60 promyelocytic leukemia cells was inhibitable by high concentrations of zinc.[19,20] Hypothermia- and hyperthermia-induced cell death in fibroblasts and human lymphoma cell lines can also be prevented by zinc.[21,22] Surprisingly, the cell death of mature T-lymphocytes and T-cell hybridoma cell death induced by IL-2 withdrawal were also inhibited by zinc, as was LPS-induced DNA fragmentation in mouse thymocytes.[23,24] Zinc at concentrations ranging from 500 to 5 μ*M* has been repeatedly found to inhibit killing by mouse cytolytic T-cells of the P815 mastocytoma, YAC-1 lymphoma, BW5147.3 thymoma, and TA-3 B-cell hybridoma cell lines.[25,26]

TABLE 2
Inhibition of Dexamethasone-Induced Cell Death in Mouse Thymocytes by Zinc and Other Standard Inhibitors of Apoptosis

Culture Conditions	Percentage of Apoptotic Cells in the Population
Cells alone (8 hr)	15.5 ± 1.9%
With Dexamethasone (0.1 μ*M*) (8 hr)	40.3 ± 2.0%
+ *$ZnSO_4$ (500 μM)*	8.2 ± 1.9%
+ Cycloheximide (50 μg/ml)	8.5 ± 1.6%
+ Actinomycin D (5 μg/ml)	9.2 ± 0.8%
+ RU38486 (1.0 μ*M*)	19.6 ± 1.1%
+ Aurintricarboxylic acid (1 m*M*)	26.6 ± 1.0%

The effectiveness of zinc as an inhibitor of apoptotic cues is demonstrated in Table 2. Thymocytes from young adult mice were incubated for 8 hr with and without dexamethasone, a glucocorticoid. It is thought that 90% of all thymocytes eventually die in the body. Thus after 8 hr, the level of apoptosis in untreated cells was about 15%, which represents the ongoing level of cell death. Conversely, dexamethasone induced about 40% of the thymocytes to die as determined by flow cytometry using the type of analysis described in Figure 1. The glucocorticoid antagonist RU38486 blocked half of all apoptotic events. Aurintricarboxylic acid, which is presumably an endonuclease inhibitor, provided only modest protection against steroid-induced death.[11] Note, however, that zinc was as effective in blocking glucocorticoid-mediated death as was cycloheximide and actinomycin D. Thus zinc, in some systems, is a very effective inhibitor of apoptotic stimuli.

However, in the case of some apoptotic stimuli, zinc provides only limited protection. Examples include zinc's protection of mouse thymocytes from adenosine-induced apoptosis and a few cases of cytolytic T-cell-induced target cell death.[16,26] In some cases, only DNA fragmentation was inhibited by the presence of zinc while other characteristics of apoptotic death emerged, such as cytoplasmic condensing of the cell, as well as blebbing of the plasma membrane. Barbieri et al. also found that zinc prevented DNA fragmentation during spontaneous apoptosis in rat thymocytes, but failed to prevent cytoplasmic collapse and loss of membrane integrity.[13] The limited ability of zinc to protect cells from apoptotic death in these systems, therefore, raises questions about whether zinc can prevent all the manifestations of apoptotic death in all types of systems.

The above studies show that high concentrations of zinc can antagonize a variety of apoptotic stimuli in a variety of cells. The mechanism(s) whereby zinc inhibits apoptotic stimuli, however, is unclear. Indeed, zinc could act at several points in the death pathway. It is also possible that the inhibitory effects of zinc may vary depending on the apoptotic stimuli used or cell type involved. Furthermore, all of these studies were carried out *in vitro*. Whether or not zinc

TABLE 3

	Total Population	Percent Apoptotic Subpopulation CD4+CD8+	CD4+CD8−	CD4−CD8+
No treatment	16.0 ± 0.2%	15.4 ± 0.6%	15.2 ± 0.5%	8.0 ± 1.2%
Zinc (100 μ*M*)	41.5 ± 1.2%	47.1 ± 1.8%	18.0 ± 0.3%	21.0 ± 0.4%

Note: Multiparameter flow cytometric analysis was used to measure the amount of apoptosis induced by incubation of mouse thymocytes in 100 μ*M* zinc which was added to RPMI-1640 culture medium supplemented with 5% fetal bovine serum. The total amount of apoptosis for the whole population is given along with the percentage of apoptosis observed in the major subpopulations found within the thymus. Values are expressed as the mean plus or minus the standard deviation of the mean for five samples. A typical thymus of a young adult mouse contained 75%, 12% and 3% respectively, of the $CD4^+CD8^+$, $CD4^+CD8^-$ and $CD4^-CD8^+$ subpopulation.

can inhibit apoptosis *in vivo* has not been addressed. Whether or not the cell experiences sufficient fluxes of intracellular zinc of high enough concentrations to prevent apoptosis is also not known. Thus, the physiological relevance of all of the above findings is also unknown at the moment.

INDUCTION OF APOPTOSIS BY MODERATE LEVELS OF ZINC

As trace element nutritionists, we remained concerned that the 0.5 m*M* to 5 m*M* concentrations of zinc salts used to inhibit apoptosis in the numerous studies discussed above might have little physiological relevance. Oddly, none of the previous investigators made any attempt to determine the level of zinc which entered the cells during the incubation periods studied. Dose-kinetic curves and tests of the ability of other metals to inhibit apoptosis were rarely examined. Further, the long-term survival of cells "protected from death" by high zinc was not determined.

For these and other reasons, we performed a series of experiments using thymocytes from young adult mice, the purpose of which was to determine if lower, more physiologically relevant levels of zinc could inhibit apoptosis. In the case of glucocorticoid-induced death, incubation of thymocytes in less than 0.3 m*M* zinc failed to provide protection against apoptosis. Furthermore, zinc had to be present when the death cue was sent. Delay of addition of zinc by as little as 2 hr began to markedly reduce the ability of the metal to block death.[27] However, to our great surprise, thymocytes incubated with zinc alone in the range of 80 to 200 μ*M* zinc sulfate experienced 30 to 40% apoptosis after 8 hr incubation (see Table 3).[27] Atomic absorption spectroscopy indicated that only about 1 μ*M* zinc actually entered the cells under these conditions. Immunophenotypic analysis showed that zinc-induced apoptosis occurred primarily in the $CD4^+$ $CD8^+$ thymocyte subset, being analogous to glucocorticoid and irradiation-induced death.[1] Additional tests showed that cells which were a product of zinc-induced death condensed in size and exhibited reduced DNA fluorescence when analyzed by flow cytometry, as did other apoptotic cells.[27] Such cells also exhibited the "ladders of DNA" upon gel electrophoresis and had the morphological characteristics of an apoptotic cell. Standard inhibitors of apoptosis such as cycloheximide and actinomycin D were able to block zinc-induced apoptosis, suggesting that some array of death genes were required. During the 8 hr incubation period, the thymocytes retained their ability to exclude trypan blue, showing no signs of necrosis. Thus, zinc at a more moderate level was also able to induce a form of death that was apoptotic in nature by a variety of classical tests.

Though zinc sulfate was used for these and many previous studies, no notable difference was observed when either zinc chloride or zinc acetate was used. Thus, the form of the salt used seems inconsequential. At similar concentrations, calcium and gold caused necrotic death. Iron, copper, and nickel failed to induce apoptosis at equivalent concentrations. However, copper at nearly 500 μM, a concentration where zinc provides protection against apoptosis, induced some apoptosis in thymocyte.[27] Though other metals should be tested, it seems evident that only a few metals may be able to induce apoptosis. At present, zinc appears to be the only trace element that at a fairly narrow range of concentrations can alternately prevent and induce apoptosis. As before, the physiological relevancy of these findings remains unknown. However, since only a fraction of the exogenously added zinc entered the cell, these experiments may be mimicking real life events after all.

MECHANISMS WHEREBY ZINC MAY INFLUENCE APOPTOTIC EVENTS

ENDONUCLEASE

Although a mechanism to explain the modulating effects of zinc on apoptotic death has not yet been defined, one of the more explored modes of action has been zinc's effects on the endogenous endonuclease associated with apoptosis-associated DNA fragmentation. Internucleosomal DNA fragmentation is believed to occur by post-translational activation of at least one Ca^{2+}/Mg^{2+}-dependent endogenous endonuclease, which cleaves DNA at the linker regions between nucleosomes.[28] Endonuclease inhibitors such as aurintricarboxylic acid can partially inhibit glucocorticoid-induced apoptosis in mouse thymocytes, as can zinc.[29] Early studies of crude nuclear lysates containing endogenous endonuclease demonstrated that calcium and magnesium could induce endonuclease activity. Conversely, at relatively "low" concentrations (5 to 50 μM), zinc completely inhibited the induction of endonuclease activity by these metals.[30] Later studies demonstrated that the chromatin in isolated thymocyte nuclei could be induced to undergo DNA fragmentation upon addition of calcium and magnesium while zinc at 5 to 50 μM was also able to inhibit these activities.[14] In 1991, Cidlowski and colleagues were able to purify the endogenous endonuclease to heterogeneity (NUC18), and confirmed that zinc acted as an inhibitor.[31,32] No putative zinc binding domain was found to exist in the protein, however, suggesting that zinc may act by competing with calcium or by changing the conformation of the endonuclease, etc. Recent observations using isolated nuclei suggest that regulation of zinc and calcium concentrations may, in turn, regulate endonuclease activity.[33,34] The fact that higher concentrations of zinc (500 to 5000 μM) are required to inhibit DNA fragmentation in intact cells than in purified preparations or isolated nuclei (5 to 50 μM) may be explained by the fact that only a very small amount of extracellular zinc actually enters the cell.

EFFECTS OF ZINC ON TRANSCRIPTIONAL ACTIVATION

The apparent ability of zinc to completely protect a variety of cell types from apoptosis suggests that zinc may be inhibiting cell death through more than one mechanism. The question of why zinc inhibits apoptotic death is, therefore, still very much open to speculation and study. In the following sections, a number of other potential targets for zinc will be reviewed for consideration as potential mechanisms for further study.

A frequent, though not universal, characteristic of apoptotic death is the requirement for *de novo* transcription and translation of genes and gene products required to complete the death process.[11] As already discussed, transcriptional and translational inhibitors effectively block many types of apoptotic death. A necessary component of apoptosis is the activation of transcriptional entities that mediate apoptosis-associated gene expression. In the case of glucocorticoid-induced apoptosis, the transcriptional activator is the glucocorticoid receptor (GR) which becomes active upon addition of ligand.[35] It occurred to us that zinc might exert an inhibitory effect on this transcriptional activation. Although the classical GR system will be used to outline the possible mechanisms below, many of these concepts can be applied to other transcriptional activators as well. We will focus on this system because of the frequency of use of glucocorticoids to induce cell death.

In the case of GR, the initial event in the pathway is the binding of a ligand or steroid to the glucocorticoid receptor in the cytoplasm. Following ligand binding, the receptor is altered such that it can translocate to the nucleus, bind to the nuclear envelope, and enter the nuclear space, presumably through nuclear pores. In the case of the docked GR, this involves the release of an hsp90 homodimer (heat shock protein) and associated proteins from the receptor core protein, which in turn exposes a nuclear or DNA recognition sequence, which indicates a zinc finger DNA binding domain. The receptor must then interact with specific response element DNA sequences to induce GR-associated transcription.[36]

Zinc might act to inhibit any one of these steps. Receptor–ligand binding and hsp90 association/dissociation have both been found to require closely spaced sulfhydryl-bearing amino acid residues in the steroid binding domain of the receptor.[37] Both processes require these thiol residues to be in a reduced form. Heavy metal ions and oxonians such as arsenite, cadmium, and selenite have all been found to prevent receptor–ligand binding in GR through crosslinking of vicinal dithiol residues. Zinc has been found to prevent thyroid receptor–ligand binding by a similar mechanism but has also been postulated to play a role in the GR-hsp90 association.[38] These results suggest that zinc might inhibit transcriptional activation at early levels of GR transduction.

Regarding subsequent events, there is evidence which also suggests that reduced thiols are necessary for interaction of the receptor with the nuclear matrix, suggesting another possible target for zinc.[39] Thus, zinc might modulate transcriptional activation at the level of the binding of transcription factor to DNA. The GR contains two zinc fingers of the Cys_2-Cys_2 variety in its DNA binding region.[40,41] The N-terminal finger has a relatively high affinity for zinc (approximately 10^{-9} *M)* due to the small number of residues separating the second and third coordinate cysteine residue. The C-terminal finger, on the other hand, has a larger number of residues separating the second and third cysteine, resulting in greater conformational strain at the coordinate binding site and a concomitant reduction in binding affinity (about 10^{-6} *M)*.[42] While the N-terminal zinc finger is probably non-exchangeable, the C-terminal finger might be more likely to lose or gain zinc depending on the intracellular zinc concentration. High concentrations of zinc such as those used to inhibit apoptosis might, therefore, enhance the DNA binding affinity. Conversely, removal of zinc from the C-terminal finger by dialysis and chromatography over Chelex-100 (a metal binding anionic exchange resin) reduces DNA binding activity in both GR.[43,44] A recent study demonstrated that the metal binding protein metallothionein introduced into a cell-free system could reduce Sp1 transcription factor binding to DNA by limiting free zinc.[45] Furthermore, alterations in dietary zinc in rats also resulted in increased metallothionein expression via specific transcription factor interaction with metal-responsive elements.[46] These results indicate that modifying the level of zinc in the environment can potentially modulate zinc finger binding affinity for DNA.[47] Increasing the extracellular zinc concentration could, therefore, modulate gene-directed apoptotic death in either a positive or negative manner at the DNA level and might explain the ability of zinc to induce and/or inhibit apoptosis over a narrow concentration range. Although this has been postulated primarily for the GR, it could apply to any apoptosis-associated transcription factor with a zinc finger binding moiety.

ZINC-INDUCED GENE EXPRESSION

Zinc and other metals have also been found to induce transcription by themselves. Heavy metals such as zinc and cadmium induce expression of a variety of genes in mammalian cells, including metallothionein-I and -II, heat shock protein 70 (hsp70), and the protooncogenes c-*jun,* c-*fos,* and, c-*myc.*[48-54] The ability of zinc to upregulate expression of the metal binding proteins metallothionein-I and -II expression in hepatocytes and lymphocytes has been particularly well-characterized. Metallothionein may regulate intracellular influxes of zinc induced by hormones and other stimuli and may also represent a stress response aimed at controlling high concentrations of potentially toxic heavy metals in cells.[55,56] It is believed that zinc and other metals induce gene expression by enhancing zinc finger transcription factor binding to metal-responsive elements (MREs), which may share transcriptional control with glucocorticoid, phorbol ester, and other enhancer regions.[49,57] Indeed, several low molecular weight metal-inducible transcription factors (MITF) have been isolated that may constitute transcription factors or cofactors. Thus, higher levels of zinc in the absence of other apoptotic stimuli might act to induce a set of death genes.

High extracellular zinc might also induce expression of metal-inducible gene products that antagonize the effects of glucocorticoids or other apoptotic stimuli. For example, zinc and cadmium induce c-*myc* expression in Swiss 3T3 cells whereas glucocorticoids downregulate c-*myc* expression in lymphocytes and lymphoid cell lines.[53,58,59] Thus, elevation of c-*myc* would seem antagonistic to glucocorticoid-induced apoptosis. Indeed, a T-cell hybridoma transfected with the c-*myc* gene under control of a non-GR-inducible promoter became glucocorticoid-resistant despite the presence of a normal GR.[60] Zinc-induced gene expression might be antagonizing GR-associated apoptosis by a mechanism of this type.

ZINC MODULATION OF APOPTOSIS-ASSOCIATED SIGNAL TRANSDUCTION

A number of cellular signal transduction events have also been found necessary for the successful initiation of apoptotic death. Signalling, normally associated with cell growth and proliferation, has been found to occur during apoptotic death, including calcium fluxes, cAMP upregulation, protein kinase C activation, protooncogene upregulation, etc.[11] In some cases, induction of the signalling event, such as increased calcium flux created by calcium ionophores, induced apoptosis in the absence of any other stimuli. Zinc may block these signalling events at one level and enhance them at another, thereby modulating apoptosis. Some potential examples are given below.

Zinc and other trace metals at high concentrations can compete with calcium for binding sites on certain calcium-binding proteins. This ability is believed to form the basis for the toxicity of many heavy metals, such as cadmium, lead, and mercury.[61] The best-characterized example of this phenomenon for zinc is the calcium sequestration/transport protein calmodulin, which normally binds four calcium ions per molecule.[62] High concentrations of zinc (on the order of 200 m*M*) inhibited *in vitro* calmodulin calcium-binding activity as measured by calmodulin-dependent Ca^{2+}-ATPase activity. This is particularly relevant to apoptotic death, since calmodulin is upregulated in thymocytes during glucocorticoid-induced apoptosis.[63] Calmodulin may also play a role in the control of intracellular calcium flux during apoptotic signalling, since calmodulin inhibitors such as trifluoroperazine and calmidazolium can inhibit glucocorticoid-induced apoptosis in mouse thymocytes.[64] Zinc might competitively inhibit the activity of calmodulin and other calcium binding proteins, resulting in the inhibition of cell death.

Zinc has been found to modulate the activity of protein kinase C (PKC) both *in vitro* and *in vivo* in mouse and human lymphocytes, both by increasing the affinity of PKC for phorbol esters and diacylglycerol and by increasing the translocation rate of the PKC–ligand complex to the plasma membrane.[65,66] PKC possesses a putative regulatory zinc binding domain, and recent work suggests that zinc and calcium are both required for interaction of PKC with diacylglycerol or phorbol ester and subsequent translocation to and interaction with the plasma membrane and the cytoskeleton in mouse lymphocytes.[67-70] Incubation of cells with high concentrations of zinc salts (10 to 100 μ*M*) increased the activity of PKC *in vivo*.[71] Thus, a regulatory mechanism for PKC implicating high concentrations of sequestered zinc has been proposed.[72-74] Zinc may also translocate within the cell upon PKC activation with phorbol esters.[75-76] Thus, at appropriate concentrations, zinc may be able to modulate apoptosis by modulating PKC activity.

CHANGES IN INOSITOL PHOSPHATE METABOLISM CREATED BY ZINC

Finally, high concentrations of zinc (500 μ*M*) have been found to induce changes in the distribution and phosphorylation state of inositol mono-, di-, tri-, and tetraphosphates isolated from T-lymphoblasts and macrophages induced to undergo apoptosis with sporodesmin and gliotoxin, respectively.[19] Inositol phosphates are phosphatidylinositol diphosphate breakdown products (along with diacylglycerol) and are involved in the regulation of intracellular calcium flux. Thus, high levels of zinc might inhibit apoptotic death at the level of inositol phosphate metabolism. The effects of more moderate levels of zinc on inositol phosphate metabolism are unknown.

ZINC-INDUCED MITOGENESIS IN LYMPHOCYTES

Perhaps one of the most intriguing uses of zinc is as a proliferation signal. Concentrations of zinc salts in the 10 to 200 μ*M* range have been repeatedly shown to have a weak mitogenic effect on mouse and human lymphocytes, both alone and in synergy with mitogenic lectins.[77,78] Zinc chloride, at 50 μ*M*, stimulates proliferation in human peripheral blood lymphocytes, though at a much lower degree than normally used mitogens, such as concanavalin A or pokeweed mitogen (approximately 10%).[79] Nevertheless, several days, not hours, are required before these mitogenic effects can be detected. However, activated cells are often more resistant to apoptotic stimuli.[1,11]

CHANGES IN CHROMATIN STRUCTURE CREATED BY ZINC

Increased zinc exerts strong stabilizing effects on chromatin *in vitro* systems, which has many implications for the initiation and inhibition of apoptosis.

In this way, zinc could indirectly affect transcriptional activation or the ability of endogenous endonuclease to cleave DNA at exposed linker regions during apoptotic death. *In vitro,* low concentrations of zinc (10 μ*M*) have been found to promote reversible denaturation and renaturation of DNA upon heating and subsequent cooling, possibly by stabilizing the DNA helix.[80] Zinc can also convert B-form DNA to Z-form at higher concentrations (1 m*M*).[81] Melting profile studies of DNA in the presence of zinc suggest that zinc may facilitate the initial transition of highly condensed chromatin to the "beads on a string" conformation resulting from the removal of H1 histones.[82] This is also supported by the observation that the presence of zinc stimulates histone phosphorylation, a post-translational modification closely coupled to the interactions of histones with DNA in the nucleosome.[83] Zinc might, therefore, relax chromatin, thereby altering the accessibility of the endonuclease to the internucleosomal linker regions. In addition, zinc at 2 m*M* was found to inhibit both the internucleosomal DNA fragmentation and associated breakdown of the nuclear lamins, and the nuclear envelope in isolated HeLa cell nuclei induced to undergo apoptosis-associated chromatin degradation.[84] However, lower concentrations of zinc (500 to 1 μ*M*) prevented changes in nuclear morphology, but did not inhibit actual internucleosomal DNA fragmentation as determined by gel electrophoresis.

ZINC EFFECTS ON MEMBRANE STRUCTURE AND MEMBRANE-ASSOCIATED PROTEINS

Zinc can also exert profound effects both on the structure of plasma membranes and on a variety of transmembrane proteins. Although the mechanism is not clear, zinc has been used as a stabilizing agent to prevent the hemolysis of erythrocytes and isolated lysosomes, making them more resistant to changes in osmolarity.[85,86] Zinc alters potassium permeability and transmembrane potential in muscle cells and across neuronal membranes.[87]

Zinc can also modify the *in vitro* activity of several membrane-associated proteins, including guanylate cyclase, adenylate cyclase, Na^+/K^+-ATPase, Ca^{2+}-dependent ATPase and microsomal NADH oxidase.[87] Although many of the signalling events in mammalian cell apoptosis remain obscure, the importance of cAMP elevation in lymphocyte apoptosis and the possible importance of G-proteins in lymphocyte signalling suggests that adenylate and/or guanylate cyclases may be important for apoptosis-associated signal transduction. Na^+/K^+- and Ca^{2+}-ATPases are also likely to be necessary for energy-dependent apoptotic death. Ca^{2+}-ATPase activity could be indirectly inhibited by zinc-associated inhibition of calmodulin activity.[62]

ROLE OF ZINC IN MICROTUBULE STRUCTURE AND ASSEMBLY

At concentrations ranging from 100 to 1 μ*M,* zinc has been repeatedly observed to stabilize microtubules both *in vitro* and *in vivo* in neurons, hepatocytes and a variety of other cell types, and prevent tubulin depolymerization even in the presence of destabilizing agents such as colchicine.[88-90] *In vitro,* zinc spontaneously induces polymerization of tubulin into sheets.[88] Agents able to induce microtubule depolymerization such as colchicine and cold shock can induce apoptosis in HL-60 cells, with chemical inhibitors of depolymerization, such as vinblastine, inhibiting cell death.[8,21] Since zinc can prevent cell death in these systems, it may also act to stabilize microtubule structure.

INHIBITION OF POLY(ADP-RIBOSYL)ATION BY ZINC

Poly(ADP-ribosyl)ation is a DNA excision-repair mechanism that may be triggered by glucocorticoid-induced apoptosis and may actually be responsible for cell death by depleting cells of their intracellular NAD and ATP stores. Shimuzu *et al.*[20] have demonstrated that the ability of 1 m*M* zinc to inhibit etoposide-induced apoptosis in HL-60 cells was accompanied by inhibition of poly(ADP-ribosyl(ation), as determined by inhibition of both mono(ADP-ribose) and poly(ADP-ribose) synthetase. Poly(ADP-ribose) synthetase possesses a zinc finger domain, suggesting that it might be regulated by zinc.[91] However, this study did not demonstrate that inhibition of poly(ADP-ribosyl(ation) was the actual cause of inhibition of cell death.

IN VIVO MODULATION OF APOPTOSIS BY SUBOPTIMAL LEVELS OF ZINC

The above review provides an almost daunting number of activities that high levels of zinc might modulate, thereby initiating apoptosis on its own or conversely blocking death processes initiated by other stimuli. However, as already discussed, the actual levels of zinc required inside the cell to modulate apoptosis are unknown, making it difficult to ascertain which of the above phenomena are most likely at work during the induction or inhibition of apoptosis by zinc. There is, however, some *in vivo* or physiological evidence that the concentration of zinc in tissues and cells may influence apoptosis. The earliest study to show that suboptimal zinc in fluids and tissues and apoptosis can be linked was the finding of apoptotic bodies in the small intestine of zinc-deficient rats by Elmes.[92] Using zinc-depleted culture medium or so-called zinc chelators, several investigators subsequently found that low cellular zinc triggered apoptosis in thymocytes and various cell lines as already discussed.[7,8] The above physiological studies also support the idea that intra- and extracellular levels of zinc could modulate apoptosis.

As was discussed, glucocorticoids are frequently used to induce apoptosis in immature cells of the immune system. The numerous sites where zinc might inhibit GR transformation and translocation were pointed out. With regard to steroids and zinc, Failla and Cousins showed that dexamethasone increased the uptake of zinc in liver cells, though the natural steroid corticosterone (CS) caused only modest accumulation of zinc in these tissues.[93] Cox also found that nonphysiological levels of CS increased zinc uptake by HeLa cells (10^{-5} *M*).[94]

A number of years ago, we showed that glucocorticoids are chronically elevated during zinc deficiency.[95,96] It is well known that zinc is redistributed during a deficiency state and this may be facilitated by Gc. However, whether or not the natural glucocorticoids affect the cellular zinc content of thymocytes, bone marrow, or other cells that could affect their propensity to undergo apoptosis is not yet clear. However, it is quite clear that the elevated glucocorticoids present in zinc deficiency and protein calorie deficiencies cause extensive apoptosis among the immature cells of the immune system. Adrenalectomy or removal of the glucocorticoids provided complete protection of the thymus, preventing it from atrophying during zinc deficiency.[96] More recently, corticosterone-containing tablets were implanted in mice which delivered chronic but low levels of glucocorticoid similar to that observed in zinc-deficient mice.[2] In a period of 10 days, the steroid caused >80% atrophy of the thymus and 50 to 80% depletion of the marrow of developing lymphoid cells of the B-lineage.[2] Clearly, malnutrition interferes with lymphopoietic processes. Flow cytometric analysis indicated that apoptotic cell death was accelerated as much as fourfold in the mice chronically exposed to glucocorticoids.[2] This correlated with the observation that several weeks of suboptimal intake of zinc also caused dramatic losses in developing B-cells in the marrow of mice.[97] Once again, mice were adrenalectomized before being placed on zinc-deficient diets. The removal of the production of glucocorticoids and their concomitant ability to induce apoptosis protected the developing lymphocytes in the marrow, as it had protected thymocytes in the thymus during zinc deficiency.[97] The evidence that apoptosis contributes significantly to the thymic atrophy and lymphopenia observed in malnourished humans and rodents is now substantial.

We have discussed numerous studies which demonstrated that moderately high and very high levels of zinc modulate apoptosis. In turn, numerous examples of biochemical processes where zinc might affect apoptotic events were also outlined. The above studies now make it evident that suboptimal levels of serum and tissue zinc created by dietary deficits can also accelerate apoptosis in the thymus and the marrow. This creates a whole new layer of questions. Does suboptimal dietary zinc have any direct effect on thymic atrophy and loss of lymphopoietic function, or are all these changes due to induction of the glucocorticoid axis that is a secondary outcome of the deficiency? If high levels of exogenous zinc can inhibit the glucocorticoid-mediated pathway of cell death, do suboptimal levels in some way enhance or increase the probability that developing lymphoid cells will enter the apoptotic state upon exposure to corticosterone or cortisol? Zinc's ability to modulate apoptosis is clearly an interesting and important finding. Determining the cellular levels of zinc that inhibit or enhance apoptosis and identifying the biochemical processes modulated by this metal that lead to induction or inhibition of apoptosis will, however, be a very difficult task.

REFERENCES

1. **Cohen, J. and Duke, R.,** Apoptosis and programmed cell death in immunity, *Ann. Rev. Immunol.,* 10, 267, 1992.
2. **Garvy, B., Telford, W., King, L., and Fraker, P. J.,** Glucocorticoids and irradiation induced apoptosis in normal murine bone marrow B-lineage lymphocytes as determined by flow cytometry, *Immunology,* 79, 270, 1993.
3. **Barry, M., Behnke, C., and Eastman, A.,** Activation of programmed cell death (apoptosis) by cisplatin, other anticancer drugs, toxins and hyperthermia, *Biochem. Pharm.,* 40, 2353, 1990.
4. **Williams, G. and Smith, C.,** Molecular regulation of apoptosis: genetic controls on cell death, *Cell,* 74, 777, 1993.
5. **Gougeon, M. and Montagnier, L.,** Apoptosis in AIDS, *Science,* 260, 1269, 1993.
6. **Telford, W., King, L., and Fraker, P. J.,** Evaluation of glucocorticoid induced DNA fragmentation in mouse thymocytes by flow cytometry, *Cell Proliferation,* 24, 447, 1991.
7. **McCabe, M. J., Jr., Jiang, S. A., and Orrenius, S.,** Chelation of intracellular zinc triggers apoptosis in mature thymocytes, *Lab. Invest.,* 69, 101, 1993.
8. **Martin, S., Mazdai, G., Strain, J., Cotter, T., and Hannigan, B.,** Programmed cell death in lymphoid and myeloid cell lines during zinc deficiency, *Clin. Exp. Immunol.,* 83, 338, 1991.
9. **Fraker, P. J., King, L., Garvy, B., and Medina, C.,** Immunopathology of zinc deficiency: a role for apoptosis, in *Human Nutrition — A Comprehensive Treatise,* Vol. 8, Klurfeld, D., Ed., Plenum Press, New York, 1993, chap. 13.
10. **Lazenik, Y., Cole, S., Cooke, C., Nelon, W., and Earn-Shaw, W.,** Nucleas events of apoptosis *in vivo* in cell free mitotic extracts. A model system for analysis of the active phase of apoptosis, *J. Cell Biol.,* 123, 7, 1993.
11. **Schwartzman, R. and Cidlowski, J.,** Apoptosis: the biochemistry and molecular biology of programmed cell death, *Endocr. Rev.,* 14, 133, 1993.
12. **Telford, W., King, L., and Fraker, P.,** Rapid quantitation of apoptosis in pure and heterogenous cell population using flow cytometry, *J. Immun. Meth.,* 172, 1, 1994.
13. **Barbieri, D. Troiano, L., Grassilli, E., Agnesini, C., and Franceschi, C.,** Inhibition of apoptosis by zinc, a reappraisal, *Biochem. Biophys. Res. Commun.,* 187, 1256, 1992.
14. **Cohen, J. J. and Duke, R. C.,** Glucocorticoid activation of a calcium-dependent endonuclease in thymocyte nuclei leads to cell death, *J. Immunol.,* 132, 38, 1984.
15. **Sellins, K. S. and Cohen, J. J.,** Gene induction by gamma-irradiation leads to DNA fragmentation in lymphocytes, *J. Immunol.,* 139, 3199, 1987.
16. **Kizaki, H., Shimada, H., Ohsaka, F., and Sakurada, J.,** Adenosine, deoxyadenosine and deoxyguanine induce DNA cleavage in mouse thymocytes, *J. Immunol.,* 141, 1652, 1988.
17. **Zheng, L. M., Zychlinsky, A., Liu, C. C., Ojcius, D. M., and Young, D. M.,** Extracellular ATP as a trigger for apoptosis in programmed cell death, *J. Cell Biol.,* 1112, 279, 1991.
18. **Flieger, D., Riethmuller, G., and Ziegler-Heitbrock, H. W. L.,** Zn^{++} inhibits both tumor necrosis factor-mediated DNA fragmentation and cytolysis, *Int. J. Cancer,* 44, 315, 1989.
19. **Waring, P., Egan, M., Braithwaite, A., Mullbacher, and Siaarda, A.,** Apoptosis induced in macrophages and T blasts by the mycotoxin sporodesmin and protection by Zn^{2+} salts, *Int. J. Pharmacol.,* 12, 445, 1990.
20. **Shimuzu, T., Kubota, M., Tanezawa, A., Sano, H., Kasai, Y., Hashimoto, H., Akiyama, Y., and Mikawa, H.,** Inhibition of both etoposide-induced DNA fragmentation and activation of poly(ADP)-ribose synthesis by zinc ion, *Biochem. Biophys. Res. Commun.,* 169, 1172, 1990.
21. **Nagle, W. A., Soloff, B. L., Moss, A. J., and Henle, K. J.,** Cultured chinese hamster cells undergo apoptosis after exposure to cold but nonfreezing temperatures, *Cryobiology,* 27, 439, 1990.

22. **Takano, Y. S., Harmon, B. V., and Kerr, J. F. R.,** Apoptosis induced by mild hyperthermia in human and murine tumour cell lines: a study using electron microscopy and DNA gel electrophoresis, *J. Pathol.,* 163, 329, 1991.
23. **Nieto, M. A. and Lopez-Rivas, A.,** IL-2 protects T lymphocytes from glucocorticoid-induced DNA fragmentation and cell death, *J. Immunol.,* 143, 4166, 1989.
24. **Thomas, D. J. and Caffrey, T. C.,** Lipopolysacharide induced double-stranded DNA fragmentation in mouse thymus: protective effect of zinc pretreatment, *Toxicology,* 68, 327, 1991.
25. **Duke, R. C., Chervenak, R., and Cohen, J. J.,** Endogenous endonuclease-induced DNA fragmentation: an early event in cell-mediated cytolysis, *Proc. Natl. Acad. Sci. U.S.A.,* 80, 6361, 1983.
26. **Zychlinsky, A., Zheng, L-M., Liu, C-C., and Young, J. D-E.,** Cytolytic lymphocytes induce both apoptosis and necrosis in target cells, *J. Immunol.,* 146, 393, 1991.
27. **Telford, W. and Fraker, P.,** Zinc-induced apoptosis in $CD4^+CD8^+abTCR^{lo}$ murine thymocytes, *J. Cell. Physiol.,* 1995, in press.
28. **Wyllie, A. H.,** Glucocorticoid-induced thymocyte apoptosis is associated with endogenous endonuclease activation, *Nature (London),* 284, 555, 1980.
29. **McConkey, D. J., Hartzell, P., Nicotera, P., and Orrenius, S.,** Calcium-activated DNA fragmentation kills immature thymocytes, *Fed. Proc.,* 3, 1843, 1989.
30. **Ishida, R., Akiyoshi, H., and Takahashi, T.,** Isolation and purification of calcium and magnesium dependent endonuclease from rat liver nuclei, *Biochem. Biophys. Res. Commun.,* 56, 703, 1974.
31. **Compton, M. M. and Cidlowski, J. A.,** Identification of a glucocorticoid-induced nuclease in thymocytes. A potential "lysis gene" product, *J. Biol. Chem.,* 262, 8288, 1987.
32. **Gaido, M. L. and Cidlowski, J. A.,** Identification, purification and characterization of a calcium-dependent endonuclease (NUC18) from apoptotic rat thymocytes. NUC18 is not histone H2B, *J. Biol. Chem.,* 266, 18580, 1991.
33. **Giannakis, C., Forbes, I. J., and Zalewski, P. D.,** $Ca^{2}+/Mg^{2+}$-dependent nuclease: tissue distribution, relationship to inter-nucleosomal DNA fragmentation and inhibition by Zn^{2+}, *Biochem. Biophys. Res. Commun.,* 181, 915, 1991.
34. **Lohmann, R. D. and Beyersmann, D.,** Cadmium and zinc mediated changes of the Ca^{2+}-dependent endonuclease in apoptosis, *Biochem. Biophys. Res. Commun.,* 190, 1097, 1993.
35. **Pratt, W. B.,** Role of heat shock proteins in steroid receptor function, in *Steroid Hormone Action: Frontiers in Molecular Biology,* Parker, M. G., Ed., Oxford University Press, London, 1993, 145.
36. **Pratt, W. B.,** Control of steroid receptor function and cytoplasmic-nuclear transport by heat shock proteins, *BioEssays,* 14, 841, 1992.
37. **Pratt, W. B.,** Transformation of glucocorticoid and progesterone receptors to the DNA-binding state, *J. Cell. Biochem.,* 35, 51, 1987.
38. **Schwartz, J. A., Mizukami, H., and Skafar, D. F.,** A metal-linked gapped zipper model is proposed for the hsp90-glucocorticoid receptor interaction, *FEBS Lett.,* 315, 109, 1993.
39. **Kaufmann, S. H., Okret, S., Wikstrom, A-C., Gustaffson, J-A., and Shaper, J. H.,** Binding of the glucocorticoid receptor to the rat liver nuclear matrix, *J. Biol. Chem.,* 261, 11962, 1986.
40. **Archer, T. K., Hager, G. L., and Omichinski, J. G.,** Sequence-specific DNA binding by glucocorticoid receptor "zinc finger peptides," *Proc. Natl. Acad. Sci. U.S.A.,* 87, 7560, 1990.
41. **Freedman, L. P.,** Anatomy of the steroid receptor zinc finger region, *Endocrine Rev.,* 13, 129, 1992.
42. **Pan, T., Freeman, L. P., and Coleman, J. E.,** Cadmium-113 NMR studies of the DNA binding domain of the mammalian glucocorticoid receptor, *Biochemistry,* 29, 9218, 1990.
43. **Freedman, L. P., Luisi, B. F., Korszun, Z. R., Basavappa, R., Sigler, P. B., and Yamamoto, K. R.,** The function and structure of the metal coordination sites within the glucocorticoid receptor DNA binding domain, *Nature (London),* 334, 543, 1988.

44. **Westin, G. and Schaffner, W.,** Biochemistry of metallothionein, *Biochemistry,* 27, 8509, 1988.
45. **Zeng, J., Heuchel, R., Schaffner, W., and Kagi, J. H. R.,** Thionein (apometallothionein) can modulate DNA binding and transcription activation by zinc finger containing factor Sp1, *FEBS Lett.,* 279, 310, 1991.
46. **Cousins, R. J. and Lee-Ambrose, L. M.,** Nuclear zinc uptake and interactions and metallothionein gene expression are influenced by dietary zinc in rats, *J. Nutr.,* 122, 56, 1992.
47. **Chesters, J. K.,** Trace element-gene interactions, *Nutr. Rev.,* 50, 217, 1992.
48. **Bremner, I. and Beatie, J. H.,** Metallothionein and the trace metals, *Ann. Rev. Nutr.,* 10, 63, 1990.
49. **Karin, M., Haslinger, A., Holtgreve, H., Richards, R. I., Krauter, P., Westphal, H. M., and Beato, M.,** Characterization of DNA seqences through which cadmium and glucocorticoid hormones induce human metallothionein-II_A gene, *Nature (London),* 308, 513, 1984.
50. **Richards, R. I., Heguy, A., and Karin, M.,** Structure and functional analysis of the human metallothionein-I_A gene: differential induction by metal ions and glucocorticoids, *Cell,* 37, 263, 1984.
51. **Mitani, K., Fujita, H., Sassa, S., and Kappas, A.,** Activation of heme oxygenase and heat shock protein 70 genes by stress in human hepatoma cells, *Biochem. Biophys. Res. Commun.,* 166, 1429, 1990.
52. **Andrews, G. K., Harding, M. A., Calvet, J. P., and Adamson, E. D.,** The heat shock response in HeLa cells is accompanied by elevated expression of the c-*fos* proto-oncogene, *Mol. Cell Biol.,* 7, 3452, 1987.
53. **Jin, P. and Ringhertz, N. R.,** Cadmium induces transcription of proto-oncogenes c-*jun* and c-*myc* in rat L6 myoblasts, *J. Biol. Chem.,* 265, 14061, 1990.
54. **Epner, D. E. and Herschman, H. R.,** Heavy metals induce expression of the TPA-inducible sequence (TIS) genes, *J. Cell. Physiol.,* 148, 68, 1991.
55. **Etzel, K. R., Shapiro, S. G., and Cousins, R. J.,** Regulation of liver metallothionein and plasma zinc by the glucocorticoid dexamethasone, *Biochem. Biophys. Res. Commun.,* 89, 1120, 1979.
56. **Cousins, R. J., Dunn, M. A., Leinart, A. S., Yedinak, K. C., and DiSilvestro, R. A.,** Coordinate regulation of zinc metabolism and metallothionein gene expression in rats, *Am. J. Physiol.,* 251, E688, 1986.
57. **Andersen, R. D., Taplitz, S. J., Wong, S., Bristol, G., Larkin, B., and Herschman, H. R.,** Metal-dependent binding of a nuclear factor to the rat metallothionein 1 gene promoter, *Mol. Cell. Biol.,* 7, 3574, 1987.
58. **Eastman-Reks, S. B. and Vedeckis, W. V.,** Glucocorticoid inhibition of c-myc, c-myb and c-Ki-ras expression in a mouse lymphoma cell line, *Cancer Res.,* 46, 2457, 1986.
59. **Yuh, Y-S. and Thompson, E. B.,** Glucocorticoid effect on oncogene/growth gene expression in human T lymphoblastic leukemia cell line CCRF-CEM, *J. Biol. Chem.,* 264, 10904, 1989.
60. **Thompson, E. B., Nazareth, L. V., Thulasi, R., Ashraf, J., Harbour, D., and Johnson, B. H.,** Glucocorticoids in malignant lymphoid cells: gene regulation and the minimum receptor fragment for lysis, *J. Steroid Biochem. Mol. Biol.,* 41, 273, 1992.
61. **Habermann, E. and Richardt, G.,** Intracellular calcium binding proteins as targets for heavy metal ions, *Trends Pharm. Sci.,* 7, 298, 1986.
62. **Brewer, G. J.,** Calmodulin, zinc and calcium in cellular and membrane regulation: an interpretive review, *Am. J. Hematol.,* 8, 231, 1980.
63. **Dowd, D. R., MacDonald, P. N., Komm, B. S., Haussler, M. R., and Miesfeld, R.,** Evidence for early induction of calmodulin gene expression in lymphocytes undergoing glucocorticoid-mediated apoptosis, *J. Biol. Chem.,* 266, 18423, 1991.
64. **McConkey, D. J., Nicotera, P., Hartzell, P., Bellomo, G., Wylie, A. H., and Orrenius, S.,** Glucocorticoids activate a suicide process in thymocytes through an elevation of cytosolic Ca^{2+} concentration, *Arch. Biochem. Biophys.,* 269, 365, 1989.

65. **Murakami, K., Whitely, M. K., and Routtenberg, A.,** Regulation of protein kinase C activity by cooperative interaction of Zn^{2+} and Ca^{2+}, *J. Biol. Chem.,* 262, 13902, 1987.
66. **Csermely, P., Szamel, M., Resch, K., and Somogyi, J.,** Zinc increases the affinity of phorbol ester receptor in T lymphocytes, *Biochem. Biophys. Res. Commun.,* 154, 578, 1988.
67. **Parker, P. J., Coussens, L., Totty, N., Rhee, L., Young, S., Chen, E., Stabel, S., Waterfield, M. D., and Ullrich, A.,** The complete primary structure of protein kinase C, the major phorbol ester receptor, *Science,* 233, 853, 1986.
68. **Csermely, P., Szamel, M., Resch, K., and Somogyi, J.,** Zinc can increase the activity of protein kinase C and contributes to its binding to plasma membranes in T lymphocytes, *J. Biol. Chem.,* 263, 6487, 1988.
69. **Zalewski, P. D., Forbes, I. J., Giannakis, C., Cowled, P. A., and Betts, W. H.,** Synergy between zinc and phorbol ester in translocation of protein kinase C to cytoskeleton, *FEBS Lett.,* 273, 131, 1990.
70. **Zalewski, P. D., Forbes, I. J., Giannakis, C., and Betts, W. H.,** Regulation of protein kinase C by Zn^{2+}-dependent interaction with actin, *Biochem. Int.,* 24, 1103, 1991.
71. **Forbes, I. J., Zalewski, P. D., Hurst, N. P., Gainnakis, C., and Whitehouse, M. W.,** Zinc increases phorbol ester receptors in intact B-cells, neutrophil polymorphs and platelets, *FEBS Lett.,* 247, 445, 1989.
72. **Forbes, I. J., Zalewski, P. D., Giannakis, C., Petkoff, H. S., and Cowled, P. A.,** Interaction between protein kinase C and regulatory ligand is enhanced by a chelatable pool of cellular zinc, *Biochim. Biophys. Acta,* 1053, 113, 1990.
73. **Zalewski, P. D., Forbes, I. J., and Giannakis, C.,** Physiological role for zinc in prevention of apoptosis (gene-directed death), *Biochem. Int.,* 24, 1093, 1990.
74. **Forbes, I. J., Zalewski, P. D., and Giannakis, C.,** Role for zinc in a cellular response mediated by protein kinase C in human B lymphocytes, *Exp. Cell. Res.,* 195, 224, 1991.
75. **Csermely, P., Gueth, S., and Somogyi, J.,** The tumor promoter tetradecanoyl-phorbol acetate (TPA) elicits the redistribution of zinc in subcelular fractions of rabbit thymocytes measured by X-ray fluorescence, *Biochem. Biophys. Res. Commun.,* 144, 863, 1987.
76. **Csermely, P. and Somogyi, J.,** Zinc as a possible mediator of signal transduction in T lymphocytes, *Acta Physiol. Hung.,* 74, 195, 1989.
77. **Chesters, J. K.,** The role of zinc ions in the transformation of lymphocytes by phytohaemagglutinin, *Biochem. J.,* 130, 133, 1972.
78. **Berger, N. A. and Skinner, A. M.,** Characterization of lymphocyte transformation induced by zinc ions, *J. Cell Biol.,* 61, 45, 1974.
79. **Ruhl, H. and Kirschner, H.,** Monocyte-dependent stimulation of human T cells by zinc, *Clin. Exp. Immunol.,* 32, 484, 1978.
80. **Borochov, N., Ausio, J., and Eisenberg, H.,** Interaction and conformational changes of chromatin with divalent cations, *Nucl. Acids Res.,* 12, 3089, 1984.
81. **Fazakerly, G. V.,** Zinc Z-DNA, *Nucl. Acids Res.,* 12, 3643, 1984.
82. **McGhee, J. D. and Felsenfeld, G.,** Nucleosome structure, *Ann. Rev. Biochem.,* 49, 1115, 1980.
83. **Sen, D. and Crothers, D. M.,** Condensation of chromatin: role of multivalent cations, *Biochemistry,* 25, 1495, 1986.
84. **Lazebnik, Y. A., Cole, S., Cooke, C. A., Nelson, W. G., and Earnshwa, W. C.,** Nuclear events of apoptosis *in vitro* in cell-free mitotic extracts: a model system for analysis of the active phase of apoptosis, *J. Cell Biol.,* 123, 7, 1993.
85. **Chvapil, M., Ryan, J. N., and Zukoski, C. F.,** The effect of zinc and other metals on the stability of lysosomes, *Proc. Soc. Exp. Biol. Med.,* 140, 642, 1972.
86. **Chvapil, M.,** New aspects in the biological role of zinc: a stabilizer of macromolecules and biological membranes, *Life Sci.,* 13, 1041, 1973.
87. **Bettger, W. J. and O'Dell, B. L.,** Minireview: a critical physiological role of zinc in the structure and function of biomembranes, *Life Sci.,* 28, 1425, 1981.

88. **Gaskin, F. and Kress, Y.,** Zinc ion-induced assembly of tubulin, *J. Biol. Chem.,* 252, 6918, 1977.
89. **Hensketh, J. E.,** Zinc stimulated microtubule assembly and evidence of zinc binding to tubulin, *Int. J. Biochem.,* 14, 983, 1982.
90. **Hensketh, J. E.,** Microtubule assembly in rat brain extracts. Further characterization of the effects of zinc on assembly and cold stability, *Int. J. Biochem.,* 16, 1331, 1984.
91. **Gradwohl, G., de Murcia, J. M., Molinete, M., Simonin, F., and de Murcia, G.,** Expression of functional zinc finger domain of human poly(ADP-ribose) polymerase in *E. coli, Nucl. Acids Res.,* 17, 7112, 1989.
92. **Elmes, M.,** Apoptosis in the small intestine of zinc deficient and fasted rats, *J. Pathol.,* 123, 219, 1977.
93. **Failla, M. and Cousins, R.,** Zinc accumulation and metabolism in primary cultures of adult rat liver cells, *BBA,* 543, 293, 1978.
94. **Cox, R. and Ruckenstein, C.,** Studies on the mechanism of hormonal stimulation of zinc uptake in human cell cultures, *J. Cell. Physiol.,* 77, 71, 1979.
95. **DePasquale-Jardieu, P. and Fraker, P. J.,** The role of corticosterone in the loss in immune function in the zinc-deficient A/J mouse, *J. Nutr.,* 109, 1847, 1979.
96. **DePasquale-Jardieu, P. and Fraker, P. J.,** Further characterization of the role of corticosterone in the loss of humoral immunity in zinc-deficient A/J mice as determined by adrenalectomy, J. Immunol., 124, 2650 (1980).
97. **Fraker, P., Osati-Ashtiani, F., Wagner, M., and King, L. L.,** Possible roles for glucocorticoids and apoptosis in the suppression of lymphopoiesis during zinc deficiency, *J. Am. Coll. Nutr.,* 1994, in press.

INDEX

A

B

D

E

F

M

N

W

Z